Praise for
REIMAGINING CAPITALISM

'This powerful and readable book is a clarion call for reimagining and remaking capitalism. The market economy, which used to generate rapid productivity growth and shared prosperity, has done much less of that over the last four decades. The shifting balance of power in favor of large companies and lobbies, the gutting of basic regulations, the increasing ability of corporations and the very rich to get their way in every domain of life, and the unwillingness of the government to step up to protect its weakest citizens are likely responsible for low productivity growth and ballooning inequality in the US economy. Rebecca Henderson argues that the market system can be reformed and this can be done without unduly harming corporations. We can have a more moral and more innovative capitalism. There is hope!'

Daron Acemoglu, co-author of *Why Nations Fail*

'If you are unsatisfied with today's economic arguments – which too often seem to present an unappealing choice between unbridled markets and old-school collectivism – you need to read Rebecca Henderson's *Reimagining Capitalism*. Henderson offers a system that rewards initiative and respects the power of free enterprise, but that also recognizes that we have a higher purpose in life than pure profit maximization. This is a book for the realist with a heart'

Arthur C. Brooks, president emeritus, American Enterprise Institute; professor of practice, Harvard Kennedy School; senior fellow, Harvard Business School; and author of *Love Your Enemies*

‘Rebecca Henderson is masterful in both elegant articulation of one of society’s great challenges and clarity of vision in laying out a roadmap for practical and essential change. *Reimagining Capitalism* is a great read, full of insights, and a refreshing perspective that is new, practical and ground-breaking, offering clear steps for transitioning to a capitalism that is both profitable as well as just and sustainable’ Mindy Lubber, CEO and president, CERES

‘Rebecca Henderson is a provocative thinker on the purpose of business in society. In her new book, she advances the dialogue about the role of business in addressing the big social and environmental challenges of our time. Hers is an important voice in an essential conversation’

Doug McMillon, president and chief executive officer, Walmart

‘In a world on fire, status quo is not a great option. Rebecca Henderson rightfully argues for a refoundation of business and capitalism and offers thought-provoking ideas on what needs to be done to address some of the world’s greatest challenges’

Hubert Joly, former chairman and CEO, Best Buy

‘A must-read for every person with a stake in our economic system since change or die is the inescapable reality confronting capitalism. The question is how. Rebecca Henderson provides investors and corporate executives with the thought leadership and compelling examples foundational for understanding how to deliver sustainable and inclusive economic growth’

Hiro Mizuno, executive managing director
and chief investment officer, GPIF

'Capitalism as we know it has gotten us this far, but to take the next steps forward as a society and species we need new ways of seeing and acting on our world. That's exactly what Rebecca Henderson's book helps us do. This is a smart, timely and much-needed reimagining of what capitalism can be'

Yancey Strickler, co-founder and former CEO, Kickstarter, and author of *This Could Be Our Future: A Manifesto for a More Generous World*

'A breakthrough book, beautifully written, combining deep humanity, sharp intellect and a thorough knowledge of business. It rigorously dismantles old arguments about why capitalism can't be transformed and will reach people who haven't yet connected with the need for deep change'

Lindsay Levin, founding partner, Leaders' Quest and Future Stewards

'With great clarity and passion, Rebecca Henderson provides a stellar guide to building a purpose-driven organization, the surest path to success in a time of rising temperatures and declining trust'

Andrew McAfee, author of *More from Less* and co-author of *The Second Machine Age* and *Machine, Platform, Crowd*

'Rebecca Henderson weaves together research and personal experience with clarity and vision, illustrating the potential for business to benefit both itself and society by leading on the most challenging issues of our day. Read, and feel hopeful'

Judith Samuelson, vice president, the Aspen Institute

'*Reimagining Capitalism* is a breath of fresh air. Written in lively prose, easily accessible to lay readers and chock full of interesting case studies, Henderson comprehensively surveys what we need to secure a workable future. Some readers may think she goes too far in places, others may think she doesn't go far enough, but everyone will want to think about the economy she urges us to create'

Larry Kramer, president of the Hewlett Foundation

'Business is at the start of a sea-change. Rebecca Henderson brilliantly captures this moment when the tide is reversing its flow, from short-term shareholder value to forward-looking common purpose. It will be an essential guide for business strategy in riding these turbulent seas'

Paul Collier, Oxford University, author of *The Future of Capitalism*

'A clarion call for business leaders to get intentional – and quickly – about purpose beyond profit and using business as a force for good. An easy read, *Reimagining Capitalism* makes the business case for those who need convincing while delivering a dose of inspiration to those of us already on the journey'

Bob Chapman, CEO of Barry-Wehmiller, author of *Everybody Matters: The Extraordinary Power of Caring for Your People Like Family*

REIMAGINING CAPITALISM

REIMAGINING CAPITALISM

How Business Can Save the World

REBECCA HENDERSON

BUSINESS

PENGUIN BUSINESS

UK | USA | Canada | Ireland | Australia
India | New Zealand | South Africa

Penguin Business is part of the Penguin Random House group of companies
whose addresses can be found at global.penguinrandomhouse.com.

First published as *Reimagining Capitalism in a World on Fire* in the United States of America by PublicAffairs, an imprint of Perseus Books, LLC, a subsidiary of Hachette Book Group, Inc. 2020
First published as *Reimagining Capitalism* in Great Britain by Penguin Business 2020
001

Printed and bound in Great Britain by Clays Ltd, Elcograf S.p.A.

A CIP catalogue record for this book is available from the British Library

HARDBACK ISBN: 978–0–241–37966–0
TRADE PAPERBACK: 978–0–241–37967–7

Follow us on LinkedIn: linkedin.com/company/penguinbusiness

www.greenpenguin.co.uk

To Jim and Harry

CONTENTS

PROLOGUE

I grew up British. The experience left (at least) two lasting marks. The first is a deep and abiding love of trees. My family life was tumultuous, and I spent much of my teens lying on the great lower limb of a massive copper beech, alternatively reading and looking up at the sky through its branches. The beech was toweringly tall—at least as tall as the three-story English manor house it stood next to—and the sun cascaded down through its leaves in greens and blues and golds. The air smelled of mown grass and fresh sunlight and two-hundred-year-old tree. I felt safe and cared for and connected to something infinitely larger than myself.

The second is a professional obsession with change. My first job out of college was working for a large consulting company, closing plants in northern England. I spent months working with firms whose roots went back hundreds of years and that had once dominated the world but were now—disastrously—failing to grapple with the challenge of foreign competition.

For many years I kept the two sides of myself quite separate. I built a career trying to understand why denial is so pervasive and change is so hard. It was a good life. I became a chaired professor at MIT and something of an expert in technology strategy and organizational change, working with organizations of all shapes and sizes as they sought to transform themselves. I spent my vacations hiking in the mountains, watching the maples burn and the aspens dance in the wind.

But I kept my job and my passions in separate boxes. Work was lucrative and fun and often hugely interesting, but it was something I did before returning to real life. Real life was cuddling on the sofa with our son. Real life was lying together on a blanket underneath the trees, introducing him to the world that I loved. I assumed that the trees were immortal: a continuously renewing stream of life that had existed for millions of years and would exist for millions more.

Then my brother—a freelance environmental journalist and the author of *The Book of Barely Imagined Beings*, a wonderful book about creatures that should not exist but do, and *A New Map of Wonders*, an intricate meditation on the physics of being human—persuaded me to read the science behind climate change. I wonder now if he was hoping to wake me up to the implications of my day job. If so, he succeeded.

It turns out that the trees are not immortal. Leaving climate change unchecked will have many consequences, but one of them will be the death of millions of trees. The baobabs of southern Africa, some of the oldest trees in the world, are dying. So are the cedars of Lebanon. In the American West, the forests are dying faster than they are growing. The comfortable assumption on which I'd based my life—that there would always be soaring trunks and the sweet smell of leaves—turned out to be something that had to be fought for, not an immutable reality. Indeed, my comfortable life was one of the reasons the forests were in danger.

And it wasn't just the trees. Climate change threatened not just my own son's but every child's future. So did rampant inequality and the accelerating tide of hatred, polarization, and mistrust. I came to believe that our singular focus on profit at any price was putting the future of the planet and everyone on it at risk.

I came close to quitting my job. Spending my days teaching MBAs, writing academic papers, and advising companies as to how to make even more money seemed beside the point. I wanted to *do* something. But what? It took me a couple of years to work out that I was already in the right place at the right time. I started working with people who had the eccentric idea that business could help save the world. A couple of them ran multibillion dollar companies. But most of them were in much smaller firms or much less exalted positions. They included aspiring entrepreneurs, consultants, financial analysts, divisional VPs, and purchasing managers. One was convinced she could use her small rug company to provide great jobs for skilled immigrants in one of the most depressed towns in New England. Several were trying to solve the climate crisis by building solar or wind companies. One was giving his life to accelerating energy conservation. One was pushing his company to educate and hire at-risk teenagers. Another was hiring convicted felons. Another was doing everything she could to clean up labor practices in the factories her firm ran across the world. Many were trying hard to channel financial capital to precisely these kinds of people: business leaders seeking to solve the great problems of our time.

All of them were skilled businesspeople, very much aware that the only way they could drive impact at scale was to ensure that doing the right thing was a "both/and" proposition—a means to both build thriving and profitable firms and to make a difference in the world. All of them were passionately purpose driven, convinced that harnessing the power of private enterprise was a hugely powerful tool to tackle problems like climate change and—perhaps—to drive broader systemic change.

I loved working with them. I still do. They strive to live fully integrated lives, refusing to wall off their work from their deepest beliefs. They struggle to create what one purpose-driven leader I know calls "truly human" organizations—firms where people are treated with dignity and respect and motivated as much by shared purpose and common values as by the search for money and power. They try to make sure that business is in service to the health of the natural and social systems on which we all depend.

But I worried. I worried that this approach to management would never become mainstream: that it was only exceptional individuals who could master the creation of both purpose and profit. I was convinced that in the long run, the only way to fix the problems that we faced was to change the rules of the game—to regulate greenhouse gas emissions and other sources of pollution so that every firm has strong incentives to do the right thing, to raise the minimum wage, to invest in education and health care, and to rebuild our institutions so that our democracies are genuinely democratic, and our public conversations are characterized by mutual respect and a shared commitment to the well-being of the whole. I couldn't see how a few purpose-driven firms could help drive the kind of systemic change that we would need to put these kinds of policies in place. My students—by this time I was teaching a course in sustainable business—shared my concerns. They had two questions: Can I really make money while doing the right thing? and Would it make a difference in the end if I could?

The book you hold in your hands is my attempt to answer these questions—the result of a fifteen-year exploration of why and how we can build a profitable, equitable, and sustainable capitalism by changing how we think about the purpose of firms, their role in society, and their relationship to government and the state.

I do not suggest that reimagining capitalism will be easy or cheap. My career has given me extensive firsthand experience of just how difficult it is to do things in new ways. For many years I worked with firms struggling to change. I worked with GM as it attempted to respond to Toyota. With Kodak, as the conventional film business collapsed in the face of digital photography. With Nokia—which at its peak sold more than half of the world's cell phones—as Apple revolutionized the business.[1] Transforming the world's firms will be hard. Transforming the world's social and political systems will be even harder. But it is eminently possible, and if you look around, you can see it happening.

I am reminded of a moment some years ago when I was in Finland, facilitating a business retreat. It was the first and last time that my agenda has included the item "5.00 pm—Sauna." Following instructions, I showed up for the sauna, took off all my clothes, and soaked up the heat. "And now," my host instructed me, "it's time to jump into the lake." I duly ran across the snow (everyone else carefully averting their eyes—the Finns are very polite about such things) and carefully climbed down a metal ladder, through the hole that had been cut in the ice, and into the lake. There was a pause. My host arrived at the top of the ladder and looked down at me. "You know," she said, "I don't think I feel like lake bathing today."

I spend a good chunk of my time now working with businesspeople who are thinking of doing things differently. They can see the need for change. They can even see a way forward. But they hesitate. They are busy. They don't feel like doing it today. It sometimes seems as if I'm still at the bottom of that ladder, looking up, waiting for others to take the risk of acting in new and sometimes uncomfortable ways. But I am hopeful. I know three things.

First, I know that this is what change feels like. Challenging the status quo is difficult—and often cold and lonely. We shouldn't be surprised that the interests that pushed climate denialism for many years are now pushing the idea that there's nothing we can do. That's how powerful incumbents always react to the prospect of change.

Second, I am sure it can be done. We have the technology and the resources to fix the problems we face. Humans are infinitely resourceful. If we decide to rebuild our institutions, build a completely circular economy, and halt the damage we are causing to the natural world, we can. In the course of World War II, the Russians moved their entire economy more than a thousand miles to the east—in less than a year. A hundred years ago, the idea that women or people with black or brown skin were just as valuable as white men would have seemed absurd. We're still fighting that battle, but you can see that we're going to win.

Last, I am convinced that we have a secret weapon. I spent twenty years of my life working with firms that were trying to transform themselves. I learned that having the right strategy was important, and that redesigning the organization was also critical. But mostly I learned that these were necessary but not sufficient conditions. The firms that mastered change were those that had a reason to do so: the ones that had a purpose greater than simply maximizing profits. People who believe that their work has a meaning beyond themselves can accomplish amazing things, and we have the opportunity to mobilize shared purpose at a global scale.

This is not easy work. It sometimes feels exactly like climbing down a metal ladder into a hole cut through foot-thick ice. But here's the thing: while taking the plunge is hard, it is also exhilarating. Doing something different makes you feel alive. Being surrounded by friends and allies, fighting to protect the things you love, makes life feel rich and often hopeful. It is worth braving the cold.

Join me. We have a world to save.

1

"WHEN THE FACTS CHANGE, I CHANGE MY MIND. WHAT DO YOU DO, SIR?"

Shareholder Value as Yesterday's Idea

> The real problem of humanity is the following: we have Paleolithic emotions; medieval institutions; and god-like technology.
>
> —E. O. WILSON

What is capitalism?

One of humanity's greatest inventions, and the greatest source of prosperity the world has ever seen?

A menace on the verge of destroying the planet and destabilizing society?

Or some combination that needs to be reimagined?

We need a systemic way to think through these questions. The best place to start is with the three great problems of our

---

The chapter title is from Paul Samuelson, who later attributed it to Keynes. "When the Facts Change, I Change My Mind. What Do You Do, Sir?" *Quote Investigator*, May 19, 2019, https://quoteinvestigator.com/2011/07/22/keynes-change-mind.

time—problems that grow more important by the day: massive environmental degradation, economic inequality, and institutional collapse.

The world is on fire. The burning of fossil fuels—the driving force of modern industrialization—is killing hundreds of thousands of people, while simultaneously destabilizing the earth's climate, acidifying the oceans, and raising sea levels.[1] Much of the world's topsoil is degraded, and demand for fresh water is outstripping supply.[2] Left unchecked, climate change will substantially reduce GDP, flood the great coastal cities, and force millions of people to migrate in search of food.[3] Insect populations are crashing and no one knows why—or what the consequences will be.[4] We are running the risk of destroying the viability of the natural systems on which we all depend.[5]

Wealth is rushing to the top. The fifty richest people among them own more than the poorer half of humanity, while more than six billion live on less than $16 a day.[6] Billions of people lack access to adequate education, health care, and the chance for a decent job, while advances in robotics and artificial intelligence (AI) threaten to throw millions out of work.[7]

The institutions that have historically held the market in balance—families, local communities, the great faith traditions, government, and even our shared sense of ourselves as a human community—are crumbling or even vilified. In many countries the increasing belief that there is no guarantee that one's children will be better off than oneself has helped to fuel violent waves of anti-minority and anti-immigrant sentiment that threaten to destabilize governments across the world. Institutions everywhere are under pressure. A new generation of authoritarian populists is taking advantage of a toxic mix of rage and alienation to consolidate power.[8]

You may wonder what these problems have to do with capitalism. After all, hasn't the world's GDP quintupled in the last fifty years, even as population has doubled? Isn't average GDP per capita now over $10,000—enough to provide every person on the planet with food, shelter, electricity, and education?[9] And, even if you think business should play an active role in attempting to solve these problems, doesn't it seem, at first glance, an unlikely idea? In the majority of our boardrooms and our MBA classrooms, the first mission of the firm is to maximize profits. This is regarded as self-evidently true. Many managers are persuaded that to claim any other goal is to risk not only betraying their fiduciary duty but also losing their job. They view issues such as climate change, inequality, and institutional collapse as "externalities," best left to governments and civil society. As a result, we have created a system in which many of the world's companies believe that it is their moral duty to do nothing for the public good.

But this mind-set is changing, and changing very fast. Partly this is because millennials are insisting that the firms they work for embrace sustainability and inclusion. When I first launched the MBA course that became "Reimagining Capitalism," there were twenty-eight students in the room. Now there are nearly three hundred, a little less than a third of the Harvard Business School class. Thousands of firms have committed themselves to a purpose larger than profitability, and nearly a third of the world's financial assets are managed with some kind of sustainability criterion. Even those at the very top of the heap are beginning to insist that things have to change. In January 2018, for example, Larry Fink, the CEO of BlackRock, the world's largest financial asset manager, sent a letter to the CEOs of all the firms in his portfolio that said the following: "Society is demanding that companies, both public and private, serve a social purpose. To prosper over time, every company must not only deliver financial performance, but also show how it makes a positive contribution to society. Companies must benefit all of their stakeholders, including shareholders, employees, customers, and the communities in which they operate."[10]

BlackRock has just under $7 trillion in assets under management, making it among the largest shareholders in every major publicly traded firm on the planet. It owns 4.6 percent of Exxon, 4.3 percent of Apple, and close to 7.0 percent of the shares of JPMorgan Chase, the world's second-largest bank.[11] For Fink to suggest that "companies must serve a social purpose" is the rough equivalent of Martin Luther nailing his ninety-five theses to Wittenberg Castle's church door.[12] The week after his letter came out, a CEO friend reached out to me to confirm that surely he didn't—really—mean it? My friend was in a state of shock. He had based a long and successful career on putting his head down and maximizing shareholder value, and to him Fink's suggestion seemed ludicrous. He couldn't imagine taking his eye off the profit ball in today's ruthlessly competitive world.

In August 2019 the Business Roundtable—an organization composed of the CEOs of many of the largest and most powerful American corporations—released a statement redefining the purpose of the corporation: "To promote an economy that serves all Americans." One hundred and eighty-one CEOs committed to lead their companies for "the benefit of all stakeholders: customers, employees, suppliers, communities, and shareholders."[13] The Council of Institutional Investors (CII)—a membership organization of asset owners or issuers that includes more than 135 public pension and other funds with more than $4 trillion in combined assets under management—was not amused, responding with a statement that said, in part:

> CII believes boards and managers need to sustain a focus on long-term shareholder value. To achieve long-term shareholder value, it is critical to respect stakeholders, but also to have clear accountability to company owners. Accountability to everyone means accountability to no one. BRT has articulated its new commitment to stakeholder governance . . . while (1) working to diminish

> shareholder rights; and (2) proposing no new mechanisms to create board and management accountability to any other stakeholder group.[14]

One of the world's largest financial managers insists that "the world needs your leadership," and some of the world's most powerful CEOs publicly commit to "stakeholder management," while many businesspeople—like my (hugely successful) CEO friend and many large investors—think they are asking for the impossible. Which of them is right? Can business really—and I mean really—rescue a world on fire?

I've spent the last fifteen years of my life working with firms that are trying to solve our environmental and social problems at scale—largely as a means of ensuring their own survival—and I've come to believe that business has not only the power and the duty to play a huge role in transforming the world but also strong economic incentives to do so. The world is changing. The firms that change with it will reap rich returns—and if we don't reimagine capitalism, we will all be significantly poorer.

I started this journey with an appropriately British degree of skepticism, but I am now surprisingly optimistic—in the "if we work really hard, we might just succeed" sense of optimistic. We have the technology and the resources to build a just and sustainable world, and doing so is squarely in the private sector's interest. It is going to be hard to make money if the major coastal cities are underwater, half the population is underemployed or working at jobs that pay less than a living wage, and democratic government has been replaced by populist oligarchs who run the world for their own benefit. Moreover, embracing a pro-social purpose beyond profit maximization and taking responsibility for the health of the natural and social systems on which we all rely not only makes good business sense but is also morally required by the same commitments to freedom and prosperity that drove our original embrace of shareholder value.

A mere decade ago the idea that business could help save the world seemed completely crazy. Now it's not only plausible but also absolutely necessary. I'm not talking about some distant utopia. It's possible to see the elements of a reimagined capitalism right now, and to see how these elements could add up to profound change—change that would not only preserve capitalism but also make the entire world better off. Indeed this book is an attempt to persuade you to give your life to the attempt.

How We Got Here

A central cause of the problems we face is the deeply held belief that a firm's only duty is to maximize "shareholder value." Milton Friedman, perhaps the most influential intellectual force in popularizing this idea, once stated that "there is one and only one social responsibility of business—to use its resources and engage in activities designed to increase its profits." From here it's not far to the idea that focusing on the long term or the public good is not only immoral and possibly illegal but also (and most critically) decidedly infeasible. It is true that the capital and product markets are ruthless places. But in its current incarnation, our focus on shareholder value maximization is an exceedingly dangerous idea, not just to the society and the planet, but also to the health of business itself. Turing Pharmaceuticals' experience with Daraprim illustrates the costs of chasing profits at the expense of everything else.

In September 2015, Turing, a small start-up with only two products, announced that it was raising the price of the generic drug Daraprim from $13.50 to $750 a tablet—an approximately 5,000 percent increase. Daraprim was widely used to treat complications from AIDS. It cost approximately $1 per pill to produce and had no competition.[15] Anyone wanting to buy Daraprim had to buy it from Turing. The move unleashed a media storm. Martin Shkreli,

Turing's CEO, was vilified in the press and accosted in public. But he was unrepentant. Asked if he would do anything differently, he replied:

> I probably would have raised prices higher. . . . I could have raised it higher and made more profits for our shareholders. Which is my primary duty. . . . No one wants to say it, no one's proud of it, but this is a capitalist society, capitalist system and capitalist rules, and my investors expect me to maximize profits, not to minimize them, or go half, or go 70 percent, but to go to 100 percent of the profit curve that we're all taught in MBA class.[16]

It's tempting to believe that Shkreli is an outlier. He is a deeply eccentric person and currently in jail for defrauding his investors.[17] But he expressed in the starkest terms the implications of the imperative to make as much money as you can, and Daraprim is not the only generic drug to have had its price hiked. In 2014, Lannett, another generic pharmaceutical producer, raised the price of Fluphenazine—a drug that is used to treat schizophrenia and is on the World Health Organization's list of most essential medicines—from $43.50 to $870—a 2,000 percent increase.[18] Valeant increased the prices of Nitropress and Isuprel—two leading heart drugs—by more than 500 percent, reportedly leaving the firm with gross margins of more than 99 percent.[19]

Surely this can't be right. Do managers really have a moral duty to exploit desperately sick people? Purdue Pharma's decision to aggressively promote the prescribing of OxyContin was—at least in the short term—hugely profitable.[20] Does this mean that it was right or even good business? Do firms have a duty to pursue the maximum possible profit, even when they know that doing so will almost certainly have significantly negative consequences for their customers, their employees, or society at large? Since December

2015, when the Paris Climate Agreement was signed, for example, the world's fossil fuel companies have spent more than a billion dollars lobbying against controls on greenhouse gas (GHG) emissions.[21] Lobbying in favor of heating up the planet may have maximized shareholder value in the short term, but in the long run, was it a good idea?

Taken literally, a single-minded focus on profit maximization would seem to require that firms not only jack up drug prices but also fish out the oceans, destabilize the climate, fight against anything that might raise labor costs—including public funding of education and health care, and (my personal favorite) attempt to rig the political process in their own favor. In the words of the cartoon: "Yes, the planet got destroyed, but for a beautiful moment in time we created a lot of value for shareholders."

Tom Toro

"Yes, the planet got destroyed, but for a beautiful moment in time we created a lot of value for shareholders."

Business was not always wired this way. Our obsession with shareholder value is relatively recent. Edwin Gay, the first dean of the Harvard Business School, suggested that the school's purpose was to educate leaders who would "make a decent profit, decently," and as late as 1981, the Business Roundtable issued a statement that said, in part: "Business and society have a symbiotic relationship: The long-term viability of the corporation depends upon its responsibility to the society of which it is a part. And the well-being of society depends upon profitable and responsible business enterprises."

A Beautiful Idea

The belief that management's only duty is to maximize shareholder value is the product of a transformation in economic thinking pioneered by Friedman and his colleagues at the University of Chicago following the Second World War. Many of their arguments were highly technical, but the intuition behind their work is straightforward.

First, they argued that free markets are perfectly efficient, and that this makes them a spectacular driver of economic prosperity. Intuitively, if every firm in an industry is ruthlessly focused on the bottom line, competition will drive all of them to be both efficient and innovative, while also preventing any single firm from dominating the market. Moreover, fully competitive markets use prices to match production to demand, which makes it possible to coordinate millions of firms to meet the tastes of billions of people. Friedman himself brought this idea to life using a very ordinary example:

> Look at this lead pencil. There's not a single person in the world who could make this pencil. Remarkable statement? Not at all. The wood from which it is made . . . comes from a tree that was cut down in the state of Washington. To cut down that tree, it took a saw. To make the saw, it took steel. To make steel, it took

> iron ore. This black center—we call it lead but it's really graphite, compressed graphite . . . comes from some mines in South America. This red top up here, this eraser, a bit of rubber, probably comes from Malaya, where the rubber tree isn't even native! It was imported from South America by some businessmen with the help of the British government. This brass ferrule? I haven't the slightest idea where it came from. Or the yellow paint! Or the paint that made the black lines. Or the glue that holds it together. Literally thousands of people co-operated to make this pencil. People who don't speak the same language, who practice different religions, who might hate one another if they ever met![22]

If Friedman were trying to make the same point today, he might use a cell phone—each of which contains hundreds of components that are manufactured all over the world.[23] But the key point is that truly competitive markets allocate resources much more effectively and much more efficiently than anything else we've tried. Indeed, pathbreaking work in the fifties and sixties established that under a number of well-defined conditions—including free competition, the absence of collusion and of private information, and the appropriate pricing of externalities—maximizing shareholder returns maximizes public welfare.[24]

The second argument behind the injunction to focus on shareholder returns rests on the normative primacy of individual freedoms, or the idea that personal, individual freedom is—or should be—the primary goal of society and that an individual's ability to make decisions about the disposition of her resources and time should be one of society's highest goals. This idea is deeply rooted in the post-Enlightenment, classical-liberal tradition of the eighteenth and nineteenth centuries. Milton Friedman and Friedrich Hayek drew from this tradition as a way to articulate an intellectual counterpoint to the Soviet Union's philosophy of centralized economic control.

Freedom, in this context, is "immunity from encroachment" or "freedom *from*"—the ability to make decisions free from the interference of others. Friedman and his colleagues suggested that free markets create individual freedom because, in contrast to planned economies, they allow people to choose what they do and how they do it and give them the resources to choose their own politics. It is difficult to be truly free when the state—or a small group of oligarchs—controls whom you work for and how much you're paid.

Third, Friedman and his colleagues argued that managers are agents for their investors. Acting as a trustworthy agent is a moral commitment in its own right, rooted in the widely shared idea that one should keep one's word and not misuse funds with which one has been entrusted. Since managers are agents, they argued, they have a duty to manage the firm as their investors would wish—which Friedman assumed would in most cases be "to make as much money as possible."

Together these three arguments make a powerful case for shareholder value maximization and are the moral force behind many businesspeople's belief that to maximize profits is to fulfill deep normative commitments. From this perspective, failing to maximize shareholder returns not only constitutes a betrayal of your responsibility to your investors but also threatens to reduce prosperity by compromising the efficiency of the system and reducing everyone's economic and political freedom. To do anything other than maximize returns—to pay employees more than the prevailing wage for no obvious benefit, for example, or to put solar panels on the roof when local coal-fired power is cheap and abundant—is not only to make society poorer and less free but also to betray your duties to your investors.

These ideas are, however, the product of a specific time and place, and of a particular set of institutional conditions. Given the realities of today's world, they are dangerously mistaken. Friedman and his colleagues first formulated them in the aftermath of the Second World War. At the time it seemed there was a serious risk that a reliance on the market would be replaced by centralized planning. Governments—after conquering economic depression and war—were popular and powerful. Capitalism was not. Enduring memories of the Great Depression that had preceded the war—at its height US GDP fell by 30 percent, while industrial production fell by almost 50 percent, and a quarter of the working population was unemployed[25]—meant that for the next twenty years, unregulated, unconstrained capitalism was regarded with suspicion nearly everywhere. This was the dominant view in Europe and in Asia. In Japan, for example, the business community explicitly embraced a model of capitalism that stressed the well-being of employees and a commitment to the long term, while in Germany, firms, banks, and unions cooperated to create a system of "co-determination" that routinely sought to balance the well-being of the firm with the well-being of employees and of the community.

This meant that for roughly thirty years after the war, in the developed world the state could be relied on to ensure that markets were reasonably competitive, that "externalities" such as pollution were properly priced or regulated, and that (nearly) everyone had the skills to participate in the market. Moreover, the experience of fighting the war created immense social cohesion. Investing in education and health, "doing the decent thing," and celebrating democracy seemed natural.

Friedman's ideas did not get much traction until the early seventies, when the turmoil of the first oil embargo ushered in a decade of stagflation and intense global competition, and the US economy came under significant pressure. Under these conditions, it was not crazy to believe that "unleashing" the market by telling managers their only job was to focus on shareholder returns would maximize both economic growth and individual freedom.

The Chicago-trained economists blamed the economy's lackluster performance on the fact that many managers were putting their own well-being before their duty to their investors. Their suggested solution—to tie executive compensation to shareholder value—was eagerly embraced by investors. Managers were told that they had a moral duty to maximize profits—indeed that to do anything else was actively immoral—and CEO pay was linked tightly to the value of the company's stock. GDP took off like a rocket and with it, shareholder value and CEO pay.[26]

But . . . meanwhile, the environmental costs of this growth—trillions of tons of greenhouse gases in the atmosphere, a poisoned ocean, and the widespread destruction of the earth's natural systems—remained largely invisible. Worldwide inequality fell as several of the developing economies—most notably China—began to catch up to Western levels of income. But in the developed world income inequality has increased enormously. The vast majority of the fruits flowing from the productivity growth of the last twenty years have gone to the top 10 percent of the income distribution, particularly in the United States and the United Kingdom.[27] Real incomes at the bottom have stagnated.[28] The populist fury that has emerged as a result is threatening the viability of our societies—and of our economies. What went wrong?

In a nutshell, markets require adult supervision. They only lead to prosperity and freedom when they are genuinely free and fair, and in the last seventy years the world has changed almost beyond recognition. Global capitalism looks less and less like the textbook model of free and fair markets on which the injunction to focus solely on profit maximization is based. Free markets only work their magic when prices reflect all available information, when there is genuine freedom of opportunity, and when the rules of the game support genuine competition. In today's world many prices are wildly out of whack, freedom of opportunity is increasingly confined to the well connected, and firms are rewriting the rules of the game in ways that maximize their own profits while simultaneously distorting the market. If firms can dump toxic waste into the river, control the political process, and get together to fix prices, free markets will not increase either aggregate wealth or individual freedom. On the contrary, they will wreck the institutions on which business itself relies.

Why Markets Are Failing Us

The Turing Pharmaceutical example illustrates the essential nature of the problem—but we can be even more precise. Markets have gone off the rails for three reasons: externalities are not properly priced, many people no longer have the skills necessary to give them genuine freedom of opportunity, and firms are increasingly able to fix the rules of the game in their own favor.

Energy is cheap because we don't pay its full costs. American consumers pay roughly five cents per kilowatt-hour (¢/kWh) for electricity from coal-fired power plants. But burning coal emits enormous quantities of CO_2 (coal is essentially fossilized carbon)—one of the leading causes of global warming. Producing a kilowatt-hour of coal-fired electricity causes at least another four cents of climate-related damage. Moreover, burning coal kills thousands of people every year and destroys the health of many more. The extraction, transportation, processing, and combustion of coal in the United States cause twenty-four thousand lives to be lost every year due to lung and heart disease (at a cost of perhaps $187.5 billion per year); eleven thousand additional lives are lost annually due to the high health burdens found in coal-mining regions (an annual cost of perhaps $74.6 billion).[29] Calculating an aggregate, global figure for the health costs associated with burning fossil fuels is enormously difficult since costs differ significantly depending on a wide range of factors, including the type of fuel and on how and where it's being burned. One estimate suggests that every ton of CO_2 emissions is associated with current health care costs of about $40, which would imply a cost per kWh of about four cents, but my colleagues who work in this area remind me that these costs can vary enormously and are often much higher.[30] When you add these costs back in, the real cost of a kilowatt-hour of coal-fired electricity is thus not 5¢ but something more like 13¢. This means we are only paying about 40 percent of the real costs of burning coal. Fossil fuel energy looks cheap—but only because we're not counting the costs we are imposing on our neighbors and on the future.

Every coal-fired plant on the planet is actively destroying value, in the sense that the costs these plants are imposing on society are greater than their total revenues, let alone their profits. For example, Peabody Energy, the largest coal company in the United States, shipped 186.7 million tons of coal in 2018 for total revenues of $5.6 billion.[31] The combined climate and health costs of burning 186.7 million tons of coal are about $30 billion, so—taking total revenue as a measure of total value creation, which is conservative—Peabody is destroying at least five times the value that it is creating.

Every time you use fossil fuels—whether it's to drive a car or to take a flight—you are creating lasting damage that you are not paying for. The production of every ton of steel, every ton of cement, and every single hamburger—to focus on a few products that are particularly energy intensive to produce—creates significant damage that isn't included in the price. The production of every cheeseburger generates approximately the same emissions as half a gallon of gasoline, and beef consumption alone is responsible for about 10 percent of global GHG emissions (and only about 2 percent of calories consumed).[32]

When you add these costs to the bottom line, it turns out that nearly every firm is causing significant damage. In 2018, for example, CEMEX, one of the largest cement companies in the world, emitted more than forty-eight million tons of CO_2—despite the fact that in 2018 about a quarter of the electricity used in its cement-producing operations was generated from renewables.[33] That's at least $4 billion worth of damage.[34] Its Earnings Before Interest, Taxes, Depreciation, and Amortization (EBITDA) that year was $2.6 billion.[35] In fiscal year 2019 the total emissions of the UK retail chain Marks & Spencer—a company that has been working hard to reduce emissions for years—were equivalent to 360,000 tons of CO_2.[36] That's about $32 million in damages. Pretax profits in the same year were £670 million.[37]

The distortion caused by the failure to price GHG emissions is enormous. Prices across the entire economy are completely out of whack. If the free market works its magic through the fact that prices capture all the information one needs to know, in this case there isn't much magic in evidence.

Markets only create genuine freedom of opportunity if everyone has the chance to play. When unchecked markets leave too many people too far behind, they destroy the freedom of opportunity that is fundamental to their own legitimacy. The world is immeasurably richer than it was fifty years ago, and inequality between countries has fallen significantly. In the 1950s half the world's population lived on less than $2 a day. Now only 13 percent live at this level, and most people have a decent subsistence.[38] But within countries inequality has jumped to levels not seen since the 1920s. In the United States and the United Kingdom, for example, the benefits of productivity growth have gone largely to the top 10 percent while real incomes have stagnated.[39]

In the United States social mobility is now significantly lower than it is in Canada and northern Europe, but it has fallen nearly everywhere.[40] The winners from the economic boom have increasingly found ways to pass on their success to their children, so that a child's success is increasingly a function of the zip code where they were born and their parents' income. Only 2–4 percent of students in the eight Ivy League schools' class of 2013 were from the bottom 20 percent of the income distribution, while between 10 and 19 percent of the class had been born to families in the top 1 percent.

A student born to the top 5 percent of the income distribution has about a 60 percent greater chance of joining the 1 percent than a student whose parents' income was in the bottom 5 percent, even if they both attended one of America's most highly regarded universities.[41] Your health is increasingly determined by your zip code. To take just one example: In 2017, the life expectancy of the residents of the poorest sections of New Bedford, Massachusetts, was slightly less than life expectancy in Botswana and Cambodia.[42]

It has also become significantly harder for entrepreneurial firms to succeed. Between 1997 and 2012, the four largest firms in every sector increased their share of their sector's revenues from 26 to 32 percent.[43] Young companies were 15 percent of the economy in 1980 but only 8 percent in 2015.[44] This increase in concentration is also reducing workers' bargaining power—and with it, both benefits and compensation—while driving up profits and prices.[45]

Markets are only free and fair if the players can't fix the rules in their own favor. In 2014, for example, two political scientists published a study exploring the relationship between popular support for a policy and the odds of it becoming law. The views of the "average citizen" in the United States, they found, don't matter at all. Proposals supported by 90 percent of the general population are no more likely to pass than proposals supported by 10 percent.[46] But if the rich wanted something done, it got done.

Spending the money to change the rules of the game in your favor can be a fantastically effective way of making money—even as it imposes significant costs on everyone else. In 1997, for example, the Walt Disney Company lobbied heavily in support of an obscure piece of legislation called the Copyright Term Extension Act (CTEA).[47]

Giving artists and authors (and filmmakers) copyright in their creations allows them to profit from their ideas—giving them the incentive to create more. But copyrights are limited so that after some reasonable period of time, other artists and authors can build on the ideas of those who have come before them. In Disney's case, for example, the movie *Snow White* is based on an old European folktale. So is *Beauty and the Beast*. The CTEA promised to extend US copyright to the life of an author plus seventy years, and to extend corporate copyrights to ninety-five years. For Disney, which was facing the risk that its most beloved—and most profitable—characters would start coming off copyright in 2023, the bill offered an additional twenty years of protection.

Disney spent slightly more than $2 million[48] lobbying for the bill—pushing so aggressively for its passage that it became laughingly known as the "Mickey Mouse Protection Act."

The bill ultimately sailed through Congress and was signed into law on October 27, 1998. My rough estimates suggest that at the time it passed, it might have been worth as much as $1.6 billion of additional income to Disney—not a bad return on slightly more than a $2 million investment.[49] There is no evidence, however, that it increased the general welfare. Rather, the reverse. Disney had argued that delaying the moment until competitors could copy its films would increase Disney's incentives to create new ones. But a group of prominent economists—including five Nobel laureates—argued that the extension had had essentially no effect on the incentives to innovate.[50] In their words, "In the case of term extension for existing works, the sizable increase in cost is not balanced to any significant degree by an improvement in incentives for creating new works."[51]

In plain language, Disney—a firm that prides itself on its wholesome family image and whose theme parks are practically a required stop for every family in the United States—had essentially laid the groundwork for charging these very families somewhere north of a billion dollars to enrich its own investors without generating anything like a comparable social benefit.

Still, this is just money. The fossil fuel companies have been pursuing a similar strategy with much graver consequences for the world. Between 2000 and 2017, the fossil fuel industry as a whole spent at least $3 billion lobbying against climate change legislation, and millions more backing groups and campaigns that denied the reality of climate change.[52]

As of this writing, Marathon Oil, the largest oil refiner in the United States, publicly acknowledges the reality of climate change and claims that it has "invested billions of dollars to make our operations more energy efficient." But it has been a vigorous supporter of the current administration's attempts to roll back existing regulations on automobile emissions, suggesting on one call to investors that the rollback could increase industry sales by 350,000 to 400,000 barrels of gasoline a day.[53] Such an increase would impose costs of between $4.3 and $4.9 billion on the rest of the world, but at a price of roughly $56/barrel would increase industry sales by between $6.9 and $7.9 billion.[54] In Washington State, oil interests outspent their opponents by two to one to defeat a measure designed to impose the first ever US carbon tax, with BP alone contributing $13 million to the effort.[55]

It's not only money that allows firms to buy favorable rules. In many situations the issues are so highly technical, so narrow, or so dull that neither the media nor the general public cares much about them. For example, changes in accounting standards are hard to understand and rarely arouse much public interest. But seemingly minor changes in accounting rules were one of the causes of the Great Crash of 2008.[56]

Profit maximization only increases prosperity and freedom when markets are genuinely free and fair. Modern capitalism is neither. If massive externalities go unpriced or uncontrolled, if true freedom of opportunity is more dream than reality, and if firms can change the rules of the game to suit themselves at the expense of the public good, maximizing shareholder value leads to ruin. Under these conditions firms have a moral duty to help build a system that supports genuinely competitive, appropriately priced markets and strong institutions. They also have a compelling economic case to do so. A world on fire threatens the viability of every business.

The Danger Ahead

For years, the proponents of the unchecked free market have been attacking government. But the alternative to strong, democratically controlled government is not the free market triumphant. The alternative is crony capitalism, or what the development economists call "extraction," a political system in which the rich and the powerful get together to run the state—and the market—for their own benefit. Extractive elites monopolize economic activity and systematically underinvest (when they invest at all) in public goods such as roads, hospitals, and schools.

There's always a trade-off. Too much focus on the public good stifles the entrepreneurial dynamic that is the lifeblood of well-functioning markets. Too much focus on economic freedom leads to the destruction of the social and natural world and to the steady degradation of the institutions that hold the market in balance.

Russia's experience illustrates this dynamic. The Soviet economy under communism grew much more slowly than the Western economies, while also greatly restricting personal and political freedoms. Following the fall of the Berlin Wall and the collapse of the Soviet empire, Russia moved aggressively to embrace a completely unconstrained market—Chicago economics in its purest form. For a golden moment it seemed as though Russia would become a developed market economy. But no one stopped to price externalities, build the institutions that would enforce the rule of law, provide decent education and health care, or ensure that firms couldn't set their own rules. Behind the smiles, the men with guns were still in charge. The Russian state sold its holdings—the vast majority of the economy—to a small group of cronies, creating a particularly nasty form of crony capitalism. The United States has a population of 327 million and a GDP of $21 trillion.[57] Russia has roughly half the population and a GDP of only $1.6 trillion.[58] Free markets need free politics: functioning institutions are great for business.

When we told the leaders of firms that their sole duty was to focus on shareholder value, we gave them permission to turn their backs on the health of the institutions that have historically balanced concentrated economic power. We told them that so long as they increased profits, it was their moral duty to pull down the institutions that constrained them—to lobby against consumer protection, to distort climate science, to break unions, and to pour money into efforts to roll back taxes and regulations. We pushed businesspeople into alliance with populist movements that actively campaigned against government, and that rejected fundamental democratic values. In the short term these alliances yielded seductive returns, but in the long term they threaten the fundamental pillars of our societies and our economies. Brexit will not be good for business. Neither will a global trade war or the end of immigration. The problem is not free markets. The problem is *uncontrolled* free markets, or the idea that we can do without government, and without shared social and moral commitments to the health of the entire society on which effective government depends.

We know what needs to be done. The United Nation's seventeen Sustainable Development Goals lay out a coherent road map—widely embraced by the business community—for building a just and sustainable world.[59] We have the technology and the brains to address our environmental problems, and we have the resources to reduce inequality. The question is not *what* should be done. The question is *how*.

Business must step up. It is immensely powerful. It has the resources, the skills, and the global reach to make an enormous difference. It also has a strong *economic* case for action. Left unchecked, global warming seems likely to shrink the American economy by roughly 10 percent by the end of the century[60] and to create almost unimaginable suffering. In the words of David Wallace-Wells, writing in *The Uninhabitable Earth* about the effects of different levels of increase in long-run average temperatures:

> Because these numbers are so small, we tend to trivialize the differences between them—one, two, four, five. . . . Human experience and memory offer no good analogy for how we should think of those thresholds, but, as with world wars, or recurrences of cancer, you don't want to see even one. At two degrees the ice sheets will begin their collapse. 400 million more people will suffer from water scarcity, major cities in the equatorial band of the planet will become unlivable, and even in the northern latitudes heat waves will kill thousands each summer. There would be 32 times as many extreme heat waves in India, and each would last five times as long, exposing 93 times more people. This is our best case scenario. At three degrees, Southern Europe would be in permanent drought and the average drought in Central America would last 19 months longer. In the Caribbean, 21 months longer. The area burned each year by wildfires would double.

By 2050 as many as a billion people could be on the move.[61] This is not a world you want to live in—and it's one that threatens the roots of our economic system. In the words of Ray Dalio, the founder of Bridgewater Associates, one of the world's largest hedge funds:

> I think that most capitalists don't know how to divide the economic pie well and most socialists don't know how to grow it well, yet we are now at a juncture in which either a) people of different ideological inclinations will work together to skillfully re-engineer the system so that the pie is both divided and grown well or b) we will have great conflict and some form of revolution that will hurt most everyone and will shrink the pie.

As Ray suggests, this is not a problem that business can solve on its own. We will only be able to tackle problems like climate change and inequality with state help—and this will require rebuilding our institutions and bringing markets and governments back in balance. Business can make an enormous difference, but only if it works together with others to build the healthy, well-run governments, vibrant democracies, and strong civil societies that will be essential to making real progress.

A reimagined capitalism—a reformed economic and political system—has five key pieces, none sufficient on its own, but each building on the other and each a vital part of a reinforcing whole. We can begin to see what this looks like in practice through the story of the transformation of a single firm.

2

REIMAGINING CAPITALISM IN PRACTICE

Welcome to the World's Most Important Conversation

Neo: I know you're out there. I can feel you now. I know that you're afraid . . . afraid of us. You're afraid of change. I don't know the future. I didn't come here to tell you how this is going to end. I came here to tell how it's going to begin. I'm going to hang up this phone, and then show these people what you don't want them to see. I'm going to show them a world without you. A world without rules or controls, borders or boundaries. A world where anything is possible. Where we go from there is a choice I leave to you.

—*THE MATRIX*, RELEASED MARCH 1999

The First Piece of the Puzzle: Creating Shared Value

In 2012, Erik Osmundsen became the CEO of Norsk Gjenvinning (NG), the largest waste handling company in Norway.[1] The waste business was an unfashionable corner of the economy, but Erik believed it was on the edge of significant transformation. Historically the business had been largely a matter of hauling garbage to local landfills. But Erik believed the future of the industry was in recycling, which had the potential to be a high-tech business selling into a global market with significant economies of scale. He also believed that the waste business held the key to addressing two of the world's great global challenges: climate change and the increasing shortage of raw materials. In his words:

> I asked myself, what other industries do we have where you can really change so much for the better? So it was the opportunity that grabbed me. I saw the potential to do something really good. The waste industry in Norway reduces Norwegian CO_2 by 7 percent, which I thought was baffling. Was that possible? We at NG collect 25 percent of all Norwegian waste and we bring 85 percent back to the industry in the form of raw materials and waste to energy. Which I thought was . . . incredible . . . I realized that our industry holds the key to achieving the circular economy—solving two global issues at the same time: the rapidly increasing global waste problem and the squeeze on the future supply of natural resources due to the projected increase in middle-class consumers around the world.

Erik was acting as interim CEO for NG and interviewing candidates for the permanent position when he made the decision to apply for the job himself. In his words:

> I remember it as if it was yesterday. It was the day before Easter and I was interviewing a really good candidate and he said look, I

> have one question for you, are you a candidate for this job? I went home, and I was thinking to myself, my God, I haven't been this engaged for decades. I went to my wife and I said, I don't know if this is a good idea and I haven't done this before at this scale. But every morning I wake up and I feel that I'm doing something that is really worthwhile and that we could actually make an impact. So after Easter I called up Reynir [the private equity partner who was acting as NG's chairman], and asked him if he could put my name in the ballot so to speak, and the rest is history.

Erik began by riding along with the waste trucks and hanging out at the depots. It quickly became clear that although the majority of employees were honest people, both NG and the industry were engaged in a range of corrupt practices. NG and its competitors were disposing of waste illegally, either by deliberately mislabeling hazardous waste as ordinary waste or knowingly dumping it into the municipal grid. It was ten times cheaper to export electronic waste to Asia illegally than to process it within Norway, while the regulations surrounding waste disposal were poorly enforced by a multitude of different authorities, and the fines for violations were tiny. One study suggested that more than 85 percent of all the waste transported in the country was in violation of the regulations.

Within NG, some managers were fudging their financials to meet short-term targets and misrepresenting the quality of the recycled materials they were selling. When Erik pushed for explanations, he was met with bemused variants of "but that's how it's always been around here." In Erik's words: "The story was always that this is the way it has always been done. Everyone else is doing it. It's always just some stupid guys in Oslo who think that things can be done differently, but we know that it can't be done differently because that won't work financially or it won't work at all."

Some people might have walked away. But Erik went back to his board, asking for the money and the time required to clean up the business. He began by putting in place a compliance policy that had to be signed by every employee. After a short amnesty he moved to a zero-tolerance regime under which infringing the policy would result in immediate termination. This was not an entirely popular move. In the first year, thirty of the top seventy line managers left the company, together with half of the senior staff. Many took their customers with them.

Erik and his team then hammered out a new vision for the company. Instead of being merely a company that hauled away waste, NG would become a global seller of industrial recycled raw materials—a global recycling powerhouse. In Erik's words, "Everything is collected. Everything is recycled. Everything is resourced. And everything is used over again as a new resource as opposed to the stuff that is dug out of mines or cut down in the forests."

He went public with what he had found, using the publicity as one lever among many to change NG's culture. He later explained:

> Hanging our dirty laundry outside the house was a very public statement not only to the industry but to our employees that we were serious. It's not lip service that we're talking here. It's not some sort of speech that you give to an industry association. We were putting our head on the block in the national media saying that we will clean things up. And we were honest about it. One of the key things we practiced from day one was this brutal truth policy.

It also gave him the opportunity to reach out to potential customers—primarily those with prominent global brands—who might be willing to pay a premium in return for peace of mind. Some customers—not as many as he had hoped, but some—responded, signing up with NG because it was the right thing to do and to avoid the possibility of scandal. Erik began to hire aggressively from firms beyond the waste management industry, looking for raw talent, new skills, and alignment with NG's new purpose. He brought in executives from as far afield as Coca-Cola, Norsk Hydro, and NorgesGruppen, Norway's largest grocery chain.

It was a costly transformation. In the first year the compliance program alone cost as much as 40 percent of NG's earnings before interest and taxes. It took several years to bring the new employees up to speed. In the meantime the local industry association threatened to expel NG for bringing the industry into disrepute, and, since Erik's agenda threatened the interests of organized crime, he himself became the target of threats.

But the new strategy also opened up unexpected opportunities. Managers who had seen the corruption firsthand and had felt powerless to do anything about it enthusiastically took up the challenge of remaking the company, and shutting the door on sloppy and illegal practices opened up space for real innovation. Slowly but surely NG began to industrialize the waste industry's value chain by embracing increasingly high-tech recycling. NG was the first Norwegian firm to purchase a state-of-the-art machine that used optical technology to sort metals. One could put an entire car in at one end and recycle 95 to 96 percent of its contents. The machine was initially rated as having a capacity of 120,000 tons a year, but within a year Erik's team was able to nearly double this number. This led in turn to a search for more waste to process, which led to a complete rethinking of the logistics of waste collection and an expansion of NG's range to all of Scandinavia. As NG stepped up its production of high-quality metals, it was able to diversify its customer base, significantly increasing the prices it received. In combination, these moves created significant economies of scale, driving down costs, increasing margins, and allowing NG to outcompete its rivals, further increasing volumes. By 2018 NG was one of the largest and most profitable waste companies in Scandinavia.

In short, Erik was able to translate his vision for improving the sustainability of the waste business into a new, highly disruptive—and highly profitable—business. The conversation around reimagining capitalism is sometimes framed in terms of a tension between profits and purpose. NG's case illustrates why this conversation is missing the point.

Business as usual is not a viable option. We have to find a different way to operate if our planet—and with it capitalism—is to survive. We need to move from a world in which environmental and social capital are essentially free—or at least someone else's business—to a world in which the need to operate within environmental limits within a thriving society is taken for granted. The transition will be massively disruptive—but like all such transitions, it will also be a source of enormous opportunity.

Everyone must breathe to live, but the purpose of living is not breathing.[2] In today's world, reimagining capitalism requires embracing the idea that while firms must be profitable if they are to thrive, their purpose must be not only to make money but also to build prosperity and freedom in the context of a livable planet and a healthy society. Erik's experience illustrates the enormous power of this kind of pro-social vision. It enabled him to create "shared value," or to build a profitable business, doing the right thing while simultaneously reducing risk, cutting costs, and increasing demand.

Contrary to what many believe, embracing pro-social goals for the firm—a pro-social purpose—is eminently legal. Nowhere in the world are firms legally required to maximize investor returns. Under US law, for example, it is probably illegal to make a business decision that will certainly destroy long-term shareholder value, but except in a few tightly defined situations such as when they have committed to sell the firm, directors have very wide latitude.[3] Under Delaware law, for example, where the majority of US companies are incorporated, directors have fiduciary duties of care, loyalty, and good faith to both the corporation *and* its shareholders. This means that directors can—and should—sometimes make decisions that do not maximize shareholder value in the short term to pursue long-term success. US directors facing hostile takeover bids do this routinely, turning down offers that value the firm at significantly more than its current stock price in the belief that the takeover will reduce the company's long-term value. They are protected by the business judgment rule, which presumes that in making a business decision, the directors of a corporation act on an informed basis, in good faith, and in the honest belief that the action taken is in the best interests of the company.

But creating shared value is not sufficient to reimagine capitalism. It's not enough to adopt a pro-social vision for the company. We also have to change the way organizations are run.

The Second Piece: Building the Purpose-Driven Organization

There are essentially two ways to run an organization. Low road firms assume that people are cogs in a machine and manage them as things, while high road firms treat people with dignity and respect, as autonomous and empowered cocreators in building a community dedicated to shared purpose. Running a high road firm might sound expensive, but it doesn't have to be. There's lots of evidence to suggest that in many circumstances high road firms are significantly more innovative and productive than their low road rivals. Making the switch from the low road to the high road is critical to reimagining capitalism for two reasons.

The first is that reimagining capitalism is not going to be easy. Deciding to create shared value is often risky. Building a just and sustainable economy will be disruptive, and the dynamics of disruption are always difficult. Purpose-inspired high road firms are much better equipped to handle the transition—as NG's example suggests—and are likely to be catalytic in driving the kinds of change we need.

The second is that building high road organizations is in itself a crucial piece of building a just and sustainable society. Not all high road firms can afford to pay higher wages, but many can, and that in itself will be a critical contribution to reducing inequality. Moreover, good jobs—jobs with meaning, in which people are treated with respect and encouraged to grow and to contribute to the best of their ability—are themselves crucial to the development of a healthy society.

Creating shared value and building high road organizations will be hugely important steps toward reimagining capitalism, but they will not be enough. Purpose-driven firms seeking to create shared value can have enormously positive impacts on the world. NG, for example, is playing a significant role in transforming the waste business. When competitors see that there is money to be made from acting in new ways, they will often embrace the change themselves. Improving energy efficiency used to be the province of inspired individuals. Now that everyone can see it's often hugely profitable, building green is fast becoming the industry-wide standard. But many firms that would like to do more find themselves constrained by the short-termism of the capital markets. Transforming the behavior of investors is just as important as transforming the behavior of firms.

The Third Piece: Rewiring Finance

Traditional finance may be the single biggest stumbling block to reimagining capitalism. As long as investors care only about maximizing their own returns, and focus only on the short term and on what can be easily measured, firms will be reluctant to take the risks inherent in seeking to exploit shared value and to embrace high road labor practices. It may be legal—it may even be morally required—to seek to address the big problems of our time, but if your investors will fire you if you do, you will leave the big problems for someone else to solve. It is essential to rewire the financial system if we are to reimagine capitalism.

Fortunately this process is already underway. If solving the big problems of our time is in the interest of investors—and in many cases it is—then the secret to persuading them to support companies seeking to do the right thing is to develop measures that demonstrate that the right thing is also the profitable thing. We need auditable, replicable metrics that capture the costs and benefits of addressing environmental and social problems so that investors too can understand the benefits of creating shared value (and so that they can hold firms to account). So-called ESG metrics—Environmental, Social, and Governance—is one response to this challenge. It took us over a hundred years to develop rigorous systems of financial accounting, and ESG metrics are still a work in progress, but they are already changing investor behavior. In 2018 more than $19 trillion—20 percent of all total financial assets under management—was invested using ESG-based information.[4]

Still, even the best metrics will not be enough to get us where we need to go. There are some things that are simply too hard to measure—and there are problems that firms could profitably solve but that would reduce investor returns if they do. A second step toward rewiring finance is to look for alternative sources of capital—to so-called impact investors, who care as much about making a difference as they do about maximizing their returns—and to consumers and employees. Consumer- and employee-owned firms are much more likely to be comfortable improving consumer and employee welfare at the expense of capital returns than conventional investors. Learning to mobilize these alternative sources of capital at scale could have powerfully catalytic effects.

Another option is to reduce the power of investors—to give managers shelter from the relentless demands of the capital markets by changing corporate governance, or the rules that specify who controls the firm. This is a tricky but exciting line of inquiry. The widespread adoption of corporate forms like the benefit corporation, could have profoundly beneficial effects—but could also have unanticipated consequences and would probably be widely resisted by today's investors.

Rewiring finance along these lines will be a critical step toward reimagining capitalism. But it will not be enough. If we can channel capital into leading-edge, purpose-driven firms, and use a focus on ESG to ensure that every firm is forced to uphold a higher standard of behavior, it would make an enormous difference. But many of the problems we face are genuinely public goods problems, and no single firm has incentives to fix them alone. We need to learn to cooperate.

The Fourth Piece: Building Cooperation

When Nike first attempted to get child labor out of its supply chain, it began by attempting to clean up its own operations, giving all its suppliers a code of conduct and auditing them regularly. This approach succeeded in improving some practices in some factories, but it proved to be impossible to fix the problem completely. Most of the large suppliers turned out to work for nearly everyone in the industry, and some of Nike's competitors had no interest in improving labor conditions—or had different ideas about how to do so. Audits proved to be a very imperfect tool for changing behavior, and many of the major suppliers routinely outsourced work to much smaller firms that proved hard to monitor. Nike was left with a serious business problem—the risk that conditions in their supply chain might cause significant brand damage—and no way to fix it.[5]

In response, Nike attempted to persuade every other major firm in the industry to join in cleaning up the entire supply chain.

Together with a number of other firms Nike pulled together the Sustainable Apparel Coalition, an organization dedicated to a cooperative response to the supply chain crisis. The core idea behind these kinds of cooperative organizations is simple—if everyone does his or her part, everyone benefits. In chocolate, for example, the major buyers of cocoa (chocolate's principle ingredient) have come to realize that the only way to ensure that cocoa is available over the long term is to band together to share the costs of building a just, sustainable supply chain.[6] In mining, the world's largest mining companies are trying to get a handle on their human rights problems by collectively agreeing to implement the UN's guiding principles on human rights.[7]

The problem with cooperating to create public goods, of course, is that even though we all benefit from their existence, we are often tempted to "free ride" by letting others do the hard work of building or maintaining them. Fortunately humans are quite good at solving public goods problems. For example, during the years my son was growing up, I hosted a large and elaborate Easter egg hunt. In the early years I attempted to give everyone lunch, but after a while my friends started to bring dishes, and the gathering gradually became a potluck affair. Lunch was usually delicious, featuring elaborate lasagnas, tasty salads, and wonderful home-baked cookies and cakes.

But a potluck only works if everyone pitches in. Taking the trouble to cook an elaborate lasagna is like taking the trouble to make sure that your suppliers are taking care of the environment and following good labor practices. There's always a temptation to free ride—to arrive with a packet of stale cookies. If everyone thinks that no one else is going to cook, no one will take the trouble, and there will be no lunch. But—particularly when everyone knows everyone else, and when everyone expects to keep working together—this rarely happens. We heap extravagant praise on the maker of the lasagna, and we punish those who bring stale cookies by teasing them unmercifully or "forgetting" to invite them back.

Sometimes—as among many families, armies, motorcycle gangs, churches, sports fans, universities, and a thousand other clubs—we become so identified with the group that we happily contribute everything we have to ensure its success. Indeed, modern psychology suggests that we are as naturally "groupish" as we are "selfish"—that humans have evolved in groups and that emotions like shame and pride and ideas like duty and honor ensure that we like being part of a team and think badly of those who loaf or take advantage.

One way to understand the history of the human race is to look at it as the story of our increasing ability to cooperate at larger and larger scales.[8] We built cooperation first within the family, then within the extended family group, and then within the village, the town, and the city. Successful nations cultivate disdain for the "other" and pride in the homeland to persuade people to pay their taxes and participate peacefully in the political process. At their best, large corporations are cooperative communities, persuading hundreds of thousands of people to work together toward a shared goal. Reimagining capitalism requires taking this ability to cooperate and mobilizing it to solve public goods problems at larger and larger scales.

The technical term for this kind of activity is "self-regulation," and it can be immensely powerful. It engages firms with each other and with third-sector and government partners in the pursuit of solutions to common problems, often prototyping solutions that prove to be a model for subsequent practice. But it is also inherently fragile. Many collaborative agreements fail to reach their goal. In the case of Nike and the textile business, for example, there continue to be firms—particularly smaller ones, or those from countries where the reputational costs of behaving badly are not that great—that are tempted to "cheat" or to buy from the lowest bidder and to tolerate questionable practices. It turns out that it's usually very hard to sustain this kind of cooperation without the help of the state—and states everywhere are failing. If we are to reimagine capitalism, we need the private sector to be part of the effort to rebuild our institutions and to fix government.

The Fifth Piece: Rebuilding Our Institutions and Fixing Our Governments

Creating shared value, learning to cooperate, and mobilizing the power of finance will all drive progress. But there are too many problems that we cannot solve without the power of government. Even if a significant fraction of America's firms adopts a high road labor strategy, it seems very unlikely that their commitment could significantly reduce inequality. Too many firms will have short-term incentives to take the low road and race to the bottom.[9] Many firms believe they simply cannot afford the cost of raising wages.

Moreover, unilaterally driving up wages is unlikely to be sustainable without moves to address the full range of factors that drive inequality in the first place, from changes in the tax code to the decline in organized labor representation, to the increasing dominance of very large firms and the failure of the US educational system to keep pace with the demands of the modern workplace. These are all issues that can only be addressed through political action. And government will only enact these kinds of measures if we can move beyond populism and gridlock. The only way we will solve the problems that we face is if we can find a way to balance the power of the market with the power of inclusive institutions, and purpose-driven businesses committed to the health of the society could play an important role in making this happen.

Business has played critical roles in building inclusive institutions in the past and could do so again. In seventeenth-century England, for example, it was a coalition of merchants and other businessmen that deposed the king and first wrote the rules of parliamentary democracy.[10] The Puritans of New England took a charter designed for a corporation and used it to build democratic government.[11]

Today's firms have enormous power to influence governments if they choose to use it. In 2015, for example, the governor of Indiana signed a bill into law that legitimized discrimination against gay people. Since today's employees will not tolerate LGBTQ discrimination, the response from the business community was swift and aggressive—and a week later the Indiana legislature backed down. Business needs to take similarly focused action in support of our institutions and our society.

Rebuilding our institutions is, of course, a collective action problem—but those firms that are seeking to create shared value, that are trying to take the high road with respect to their employees, and that are learning to act cooperatively are ideally positioned to solve this problem. They have committed themselves to making a difference, and they are finding that in many cases they can only reach their goals with the support of governments firmly committed to the public good.

REBUILDING OUR INSTITUTIONS requires the development of new ways of behaving and new ways of believing, just as much as it requires the development of new laws and new regulations.

We will not reimagine capitalism unless we rediscover the values on which capitalism has always been based, and have the courage and the skill to integrate them into the day-to-day fabric of business. To pretend that this is not the case is to critically misrepresent the truth of our current situation. We are destroying the world and the social fabric in the service of a quick buck, and we need to move beyond the simple maximization of shareholder value before we bring the whole system crashing down around our heads.

I'm often tempted to downplay the role that the courage to express one's personal values will play in driving the necessary changes. Sometimes when I'm standing on a stage in full regalia (stylish black jacket, colorful scarf, the highest heels I can manage) in front of a room full of powerful people, I'm tempted to tell them that they should try to solve the world's problems simply because it will make them all more money. It has the great virtue of being true, and I know they'll love it. I sometimes get concerned that if I start talking about "values" and "purpose," they will write me off as a simpering female who doesn't get the hard realities of life in the business world. But change is hard. I spent the first twenty years of my career trying to persuade firms like Kodak and Nokia to change their ways, and I know there are always a thousand reasons to put one's head down, ignore what's coming, and focus on next quarter's results.

I will never forget a conversation I had once at Motorola's paging division. It was a hot day in Florida, and I was in a windowless conference room, holding a rough prototype of something that looked very much like a smartphone, several years before anyone had heard of a Blackberry—let alone an iPhone. I had been preaching the benefits of making a significant investment in the new technology, but the divisional manager looked at me skeptically—to this day I remember the curve of his eyebrow—and said:

> I see. You're suggesting that we invest millions of dollars in a market that may or may not exist but that is certainly smaller than our existing market, to develop a product that customers may or may not want, using a business model that will almost certainly give us lower margins than our existing product lines. You're warning us that we'll run into serious organizational problems as we make this investment, and our current business is screaming for resources. Tell me again just why we should do this?

New ways of doing things nearly always look profoundly uncertain and are nearly always less profitable than existing ways of behaving. But grasping them often yields rich rewards, while denying them—as Motorola did—often leads to disaster. Twenty years of research have taught me that the firms that were able to change were those that had a reason to do so. Purpose is the fuel that provides the vision and the courage that is required to reimagine capitalism.

Running a company that's trying to make a difference in the world is not for the faint of heart. The successful purpose-driven leaders I know are almost schizophrenic in their ability to switch from a ruthless focus on the bottom line to passionately advocating for the greater good. Hamdi Ulukaya, the founder and CEO of Chobani, and one of the most authentically purpose-driven leaders out there, is effectively two people: a driven businessman and a compassionate humanitarian. "I'm a shepherd and I'm a warrior," he says when asked. "I come and go between those two. I'm a nomad, and nomads are the most real people. You can't pretend."

For several years I had the honor of facilitating Paul Polman's strategic retreats. Paul was the CEO of Unilever at the time, and he was in the midst of trying to persuade his senior team that committing the organization to solving the world's problems was not only the right thing to do but also the royal road to industry leadership. He would move seamlessly from passionately discussing the thousand ways in which Unilever could make the world a better place to ruthlessly cross-examining one of his divisional presidents about why she had missed her sales targets for the quarter and what exactly she was going to do about it—without missing a beat.

Running a company committed to doing the right thing is harder than running a conventional company. It's about being able to be a superb manager *and* a visionary leader. About being ruthlessly focused on the numbers and simultaneously open to the wider world. But it is eminently possible—and a lot more fun—to manage this way. Leaders like Hamdi and Paul are reimagining capitalism. They are creating value for their investors, while never losing sight of their responsibility to the world on which they depend. Building a just and sustainable world will not be easy or cheap. But in my view we have no realistic alternatives. We must find a way to make this work.

A CEO with whom I work recently described a conversation he'd had with two of his largest investors:

> I gave them the usual spiel about how our operating margins were up and how the investments we'd been making for growth were paying off, and they asked me the usual questions. Then I asked them if they thought climate change was real and, if it was, if the world's governments were going to fix it. Yes, they said—and no, governments weren't going to fix it. There was a pause. I asked them if they had children. They did. So I said, "If government isn't going to fix it, who will?" There was another pause. Then we started a real conversation.

Welcome to the world's most important conversation.

3

THE BUSINESS CASE FOR REIMAGINING CAPITALISM

Reducing Risk, Increasing Demand, Cutting Costs

Money is like love; it kills slowly and painfully the one who withholds it, and enlivens the other who turns it on his fellow human.

—KAHLIL GIBRAN, "YESTERDAY & TODAY, XII," FROM *KAHLIL GIBRAN'S LITTLE BOOK OF LIFE*

Money is, in some respects, like fire. It is a very excellent servant, but a terrible master.

—P. T. BARNUM, *THE ART OF MONEY GETTING: OR GOLDEN RULES FOR MAKING MONEY*, 1880

Having conversations about important issues is, of course, important. But is there evidence that there is a business case for creating shared value or for treating people well and reducing environmental damage? Definitely.

Thousands of firms are even now making billions of dollars while simultaneously addressing social and environmental problems. In the United States, solar is now an $84 billion industry and employs more people than coal, nuclear, and wind combined.[1] Wind provides 7 percent of US electricity.[2] India just canceled fourteen large coal-fired power stations because the price of solar energy has fallen so dramatically.[3] Two million plug-in electric vehicles were sold last year, and sales are growing exponentially.[4] The market for meats is expected to be a $140 billion industry within the next ten years.[5]

But to reimagine capitalism we have to reimagine every business—not just the sexy ones. Companies everywhere are claiming to make a difference because millennials won't work for them otherwise. But is there actually a business model for going green? Norsk Gjenvinning is a great story—but, well, it is in the waste business after all. So it's perhaps not that surprising to learn that one could make money by doing a better job of recycling. And Erik Osmundsen is the CEO, and has the authority to push his agenda. But can people who have regular jobs—who face the daily pressures of competition and the skeptical eyes of peers and bosses—really find a way to reimagine capitalism? Even if they sell, say, teabag tea? The answer is often yes—but doing so requires allies, courage, and organizational savvy.

Hunting for Reasons to Do the Right Thing

Michiel Leijnse joined Unilever in the summer of 2006 as the brand development manager for Lipton tea. He had come from Ben & Jerry's ice cream, a small company where a strong brand name combined with a super-premium product had allowed the company to develop a famously green supply chain—and to charge for it. But Lipton, the largest tea brand in the world, was a very different proposition.[6]

Tea is the world's most popular drink after water. Nearly half of humanity drinks it every day, and in 2018 the world imbibed about 273 billion liters of the stuff—or about a trillion cups.[7] Unilever sells just under $6 billion worth of tea every year, most of it in teabags.[8] Selling teabag tea is a highly competitive business. Teabags are cheap—as of this writing, for example, Walmart will sell you a hundred Lipton teabags for $3.48, or for the princely price of 3.5¢ each—and most consumers don't see much difference in quality or taste between the big brands.[9] When Michiel took over, it looked as if the industry might be in a death spiral. Chronic oversupply, coupled with a lack of any real differentiation between products, was driving the big brands to cut prices—which put further pressure on everyone to cut prices even further. In 2006 the price of tea was less than half of its mid-1980s peak. What was to be done?

In response, Michiel—working closely with colleagues from across the tea business—came up with a startlingly counterintuitive proposal. They recommended that Unilever should publicly commit to purchasing 100 percent sustainably grown tea. This would be a massive undertaking—requiring, among other things, training over half a million smallholder farmers—and it would significantly raise the price that the firm paid for tea. In other words, Michiel was proposing to raise his costs in an intensely competitive business in the middle of an ongoing price war. To say that this was not a textbook move is an understatement. If Michiel had walked into my office and asked for my advice, I might well have told him to lie down until the feeling went away. In the event, it took Michiel and his colleagues five months of one-on-one conversations to persuade their bosses that they hadn't lost their minds.

What were they thinking?

Michiel and his colleagues made several arguments. The first was around ensuring supply. Growing tea can be a dirty business. For some smallholders, tea production implies the conversion of tropical forests into agricultural land, which can lead to reductions in biological diversity and to soil degradation.[10] Logging for the firewood needed to dry tea can lead to local deforestation, which in turn can lead to problems in water retention. However, for most farmers unsustainable practices are the result of a focus on increasing yields rather than on increasing acreage. Conventional tea production entails the large-scale application of insecticides, pesticides, and fertilizers. Together they reduce soil quality and increase erosion. Years of commoditization have contributed to a downward price spiral that puts pressure on workers and the environment as farmers try to safeguard their income. Tea growers struggle to maintain production on degraded, eroding soils using ever more chemicals, further accelerating erosion and soil degradation. Tea production is also particularly vulnerable to global warming, as increasing temperatures, droughts, and floods are all likely to make growing tea more expensive and more difficult.[11] The team suggested that current practices were putting the entire viability of the supply chain at risk. In Michiel's words: "If we didn't do something to transform the industry, at some point we just wouldn't be able to get the quality and quantity of tea we needed." Since Unilever buys a significant fraction of branded global tea supplies, this was a material risk to the business. In the case of cocoa, for example, the key ingredient in chocolate, unsustainable growing practices coupled with the impact of climate change have meant that supplies have lagged significantly behind worldwide demand, and cocoa prices have become increasingly volatile as a result.[12]

The second set of arguments focused on the need to protect Unilever's tea brands. Working conditions on conventional tea plantations can be grim. Tea harvesting is a labor-intensive business, requiring workers to pick the top two to three leaves of the tea plants every ten to twelve days. But tea workers are often paid less than $1 a day, and many suffer from inadequate housing and sanitation, and have little or no access to health care or education for their children. In Bangladesh and India, tea workers routinely suffer from acute malnutrition and are among the poorest of all workers.[13] The team argued that in failing to insist on better practices in its supply chain, Unilever ran the risk of being attacked—and that in a mass media age, these kinds of attacks could be immensely costly.

Team members further suggested that it would be possible to persuade tea suppliers to embrace more sustainable practices. Michiel's colleagues had dragged him to Kericho—a beautiful twenty-one-thousand-acre Kenyan tea plantation that Unilever had owned for many years. Tea production at Kericho was significantly more sustainable than it was in the rest of the industry. For example, tea bush prunings were left on the field to rot, rather than being removed as waste or for use as firewood or cattle food, a practice that maximized soil fertility and water retention. The estate carefully managed its fertilizer use. On-site hydropower provided reliable electricity at one-third the cost of power bought from the Kenyan grid, and the tea was dried using wood sourced from fast-growing eucalyptus forests planted on the edge of the estate. Kericho made only minimal use of agrochemicals and other pesticides, both because of the favorable climate and also through appropriate management of the surrounding land, which was home to the natural predators of many pests.

At the same time Kericho was achieving some of the highest yields in the world, with tea production of 3.5 to 4.0 tons per hectare, nearly double that of most conventional tea plantations. That meant that they could afford to pay their sixteen-thousand-plus employees more than two and a half times the local agricultural minimum wage. In addition, employees had free access to company housing and health care, and employee children were educated in company-owned schools.[14] If Unilever could come up with a way to cover the costs of training their suppliers and of certifying the tea, it seemed plausible that the suppliers would be willing to switch—and that Unilever would only have to pay about a 5 percent premium for the tea.

The third—and most crucial—argument that Michiel and his colleagues made was that embracing sustainability would increase consumer demand for Unilever's teas. They didn't think they had any chance of persuading Unilever's tea-drinking customers to pay more for their tea. When asked, most consumers talk a good game. In one recent global study, for example, nearly three-quarters of consumers claimed that they would change their consumption habits to reduce their impact on the environment.[15] Nearly half claimed to be willing to forgo a popular brand name for an environmentally friendly product.[16] In Latin America and in Africa and the Middle East, roughly 90 percent of respondents expressed an urgent need for companies to embrace environmental issues.[17] But by and large most consumers, most of the time, won't pay more for sustainable products. Middle-aged, middle-class women will pay more for sustainable products in some settings, and some people will pay more for high-end products like coffee and chocolate, but for most people a product's sustainability is—at least today—something that is a "nice to have" rather than a "must have."[18]

A 5 percent increase in the cost of your most important raw material is a lot of money in the midst of a price war—particularly if you don't think you can raise prices—but at Ben & Jerry's, Michiel had helped to launch the world's first Fair Trade ice cream, and he had become particularly attuned to the ways in which concern over the environment and labor practices was beginning to shape buying behavior. He hoped that at least some of Lipton's consumers were sufficiently concerned that they would be willing to switch to a sustainable brand.

By this stage you can probably see why it took Michiel and the team nearly six months to persuade senior management to approve the idea. But they did. My sense—based on interviews conducted several years later—is that the team was passionate about the fact that this was the right thing to do, and reasonably convincing that at least one of the three business models they had argued for might pay off. Unilever had long been a values-driven company, and my guess is that it was the combination of purpose and plausible economics that clinched the deal.

In any event, there turned out to be an element of truth in all three arguments. The Unilever estates in Kenya and Tanzania were the first sites to be certified as growing sustainable tea. The team then identified a priority list of its larger suppliers in Africa, Argentina, and Indonesia. Many of these estates were already professionally managed and were certified following adjustments to existing practices, using available tools.[19] The next step was the certification of the five hundred thousand Kenyan smallholders from whom Unilever purchased tea. Working with the Kenyan Tea Development Agency (KTDA) and the IDH, the Dutch Sustainable Trade Initiative, Unilever designed a program that "trained the trainers" and led to the rapid diffusion of sustainable farming practices across the country. Each tea factory elected thirty to forty lead farmers, each of whom received approximately three days of training. Each lead farmer was in turn expected to train approximately three hundred other farmers through farmer field schools, with the focus of the training being hands-on demonstrations of sustainable agricultural practices. Most of the new techniques did not require huge changes in practice or much investment. For example, getting farmers to leave their prunings in the field (to improve soil quality) rather than removing it for use as firewood required persuading farmers to plant trees for fuel. Tree seeds were very cheap, and Unilever subsidized the cost. Farmers were also encouraged to make compost from organic waste rather than burning it, as well as to make better use of waste and water.

Some changes were expensive. For example, the certification standards required the use of personal protective equipment for the spraying of (approved) pesticides. This could cost up to $30, half a month's salary for a smallholder.[20] In these cases, the Rainforest Alliance—the NGO in charge of certification—worked with organizations like Root Capital and the International Finance Corporation to assist with the purchase, and in some places, the local smallholders pooled money to buy a single set, which was shared.[21] One study suggested that total net investments were less than 1 percent of total cash farm income for the first year.[22]

Many of the farms saw yield gains of 5 to 15 percent from the implementation of more sustainable practices, as well as improvements in the quality of the tea, reductions in operating costs, and the chance to realize higher prices. Average income increased by an estimated 10 to 15 percent.[23] But according to Richard Fairburn, the manager of the Kericho estate, the most salient benefit to farmers was more intangible: "The Kenyan smallholders are ultimately interested in creating a farm in good health that can be passed on to future generations. That was the 'sustainability' that resonated with them."

What about the promised benefits to Unilever?

By 2010 all Lipton Yellow Label and PG tips tea bags in western Europe, Australia, and Japan were fully certified, and by 2015 all the tea in Lipton tea bags—approximately a third of Unilever's tea volume—came from Rainforest Alliance–certified estates. The effort had changed the lives of hundreds of thousands of tea workers and demonstrated that it was possible to significantly increase the health and resilience of the supply chain. But, as forecasted, Unilever's costs had significantly increased. Tea supplies were still strong, and none of Unilever's brands had been hit by unmanageably bad publicity, but one of the problems with risk avoidance as a business case is that it's often quite hard to measure.[24] Michiel needed to demonstrate an increase in demand.

Nothing could be accomplished on this front unless his colleagues on the ground in each market could be persuaded to put marketing muscle behind the project, and not all of them were convinced it was a good idea to make the attempt. At the time, marketing at Unilever was a highly decentralized activity, with each country having its own marketing team, and each making its own decisions about whether and how much to push sustainability as a part of Lipton's identity. In the first year of the campaign, at least one major region—the United States—decided against using the sustainability theme in its marketing campaigns, and another—France—was highly skeptical that it would have an effect, and only introduced it under pressure. But in those markets where the local organization enthusiastically embraced the idea, Unilever's share increased significantly. The UK market, for example, was roughly 10 percent of Unilever's tea sales and dominated by two major brands—Unilever's PG tips and its rival, Tetley Tea. Each had roughly a quarter of the market.[25]

The PG tips brand was a mass-market, working-class brand, and its advertising campaigns were infused with offbeat British humor.[26] The marketing team treated the sustainability initiative as a major brand innovation and devoted their entire year's €12 million (approximately $13/£10 million) marketing budget to promoting the effort. The challenge was to find a message that would resonate with its consumers, while maintaining consistency with the brand's core proposition. "It was a huge challenge," explained one member of the campaign. "We had to talk to mainstream consumers in a way that explained a complex topic without preaching, all in a language aligned with the brand." The chosen message, "Do your bit: put the kettle on," emphasized the positive action that consumers could take by drinking PG tips. The campaign tried to keep the lighthearted spirit of the brand's previous campaigns and used its well-established characters: a talking monkey called Monkey and a working-class man named Al. In one of the ads, for example, Monkey, presenting a slideshow in the kitchen, explained to Al what becoming sustainable meant, and how easy it was for him to do the right thing.

Prior to the campaign, PG tips and Tetley Tea had been battling hard for the top spot in the British market. But following the campaign, PG tips saw its market share increase by 1.8 points, while Tetley remained relatively flat. Repeat purchase rates increased from 44 to 49 percent, and sales of PG tips increased by 6 percent. Surveys suggested that there had been a steady increase in the perception of PG tips as an ethical brand following the launch of the campaign.

In Australia, the Lipton brand held nearly a quarter of the €260 million[27] ($288/A$345 million) market. The local team chose the phrase "Make a Better Choice with Lipton, the world's first Rainforest Alliance Certified tea," and introduced a €1.1 million ($1.2/A$1.4 million) campaign. Sales increased by 11 percent, and Lipton's market share rose from 24.2 to 25.8 percent. In Italy, where Unilever had an approximately 12.0 percent share,[28] the message chosen was "your small cup can make a big difference," and sales increased by 10.5 percent.

In the context of a ruthlessly competitive consumer goods business like tea, these are great numbers, and my back-of-the-envelope calculations suggest that Unilever broke even on its investments within the first few years, while simultaneously significantly strengthening its brands. In 2010 it was one of the experiences that led Paul Polman, the incoming CEO, to commit the firm to a "Sustainable Living Plan," a plan that set wide-ranging company-wide goals for improving the health and well-being of consumers, reducing environmental impact, and, perhaps most ambitiously, sourcing 100 percent of agricultural raw materials sustainably by 2020. This goal implied massive transformation in a supply chain that sourced close to eight million tons of commodities across fifty different crops, and Paul's belief that it could be a source of competitive advantage was rooted—at least in part—in Unilever's experience in tea. Michiel and his colleagues had demonstrated that a commitment to sustainability could pay off—or that, at least for Unilever, it was possible to create shared value at a billion dollar scale.

Michiel's success highlights two of the four available pathways for creating shared value: reducing risk and increasing demand. I'll explore risk reduction as a path to driving change below. Using sustainability to increase demand at the margin is increasingly widespread. Consumers won't usually pay more for sustainability. But if they find a product that they like—one that ticks all the right boxes in terms of quality, price, and functionality—then many of them will switch to the more sustainable product. In June 2019, Unilever announced that its purpose-led, "sustainable living" brands were growing 69 percent faster than the rest of the business and generating 75 percent of the company's growth.[29]

Michiel's success also demonstrates the ways in which reimagining capitalism is not just a game for CEOs. Michiel found allies at Lipton, particularly among managers in the supply chain who had spent years on the ground in Africa and India and who were passionately committed to changing the way the tea business was run. Together they found a way to build and implement a business case that helped trigger the transformation of the entire company.

Using the embrace of shared value to reduce risk and increase demand is a powerful way to create economic returns. Walmart's culture-changing experience with Hurricane Katrina[30] led it to the discovery of another great reason to embrace sustainability: the fact that there turns out to be money—a great deal of money—lying on the floor. Cleaning up one's environmental footprint can be a great way to cut costs.

Walmart and the $20 Billion Bill

You would not have guessed from his background that Lee Scott was going to grow up to be a passionate environmentalist. Scott grew up in Baxter Springs, Kansas, where his father owned a Phillip's 66 gas station and his mother was a music teacher at the local elementary school. After graduating from high school, he went to work at a local company that made tire molds. By the time he was twenty-one, he was working the night shift to pay for college and living with his wife and son in a tiny trailer.[31]

Seven years later he was living in Springdale, Arkansas, working as a terminal manager for a trucking company called Yellow Freight. Here—while trying to collect a debt for Yellow Freight—Scott met David Glass, who a decade later was going to become Walmart's second CEO. Glass refused to pay the bill, believing it to be erroneous, but—impressed by Scott's sincerity and drive—offered Scott a job. Scott refused, later remarking that he told himself, "I'm not going to leave the fastest-growing trucking company in America to go to work for a company that can't pay a $7,000 bill!" But two years later, Glass succeeded in persuading him to join Walmart as assistant director of logistics, and twenty years later, Scott became Walmart's third CEO.[32]

It was a tough time for Walmart. Scott found himself in the midst of a media storm. By all the traditional measures, Walmart was an outstandingly successful company—indeed in many ways it symbolized all that was best about free market capitalism. It exemplified the way that "outsiders" could make it big: Walmart was founded in rural Arkansas on the basis of the radically unlikely idea that providing retail service to rural America could be a profitable business. Over thirty years Walmart remade retailing, developing skills in logistics, purchasing, and distribution that led to it becoming one of the largest companies in the world. In 2000, when Scott took over as CEO, Walmart had approximately $180 billion in revenue and employed more than 1.1 million people.[33]

While outsiders focused on Walmart's stunning financial returns, insiders like Scott were just as excited about the impact that Walmart had on people's lives. If you worked at Walmart, "save money, live better" was not an empty corporate slogan, but a compelling statement of the deepest purpose of the firm. One independent study found that between 1985 and 2004, Walmart saved the average American household $2,329 per family, or about $895 per person.[34] Seventy-five percent of Walmart's management team had come up through the ranks, and Scott and his colleagues saw the firm as a powerful engine of economic advancement for people who might otherwise be excluded from the economic mainstream.

But critics of the company had a very different view of its impact, and as the 2000s unfolded, Walmart found itself increasingly under fire.[35] Walmart was accused of hurting downtown areas by driving out smaller independent stores that were unable to compete with Walmart prices. Unions claimed that Walmart crossed the line when it came to anti-union activities, and that its wages and employment practices forced large numbers of Walmart workers into government-support programs to supplement their food, rent, and medical care. The company was sued for gender discrimination and investigated for employing workers living and working illegally in the country. It was accused of violating child labor laws and buying products from suppliers that were made by child labor. Walmart's competitors and suppliers faced similar allegations, but Walmart's conduct received significantly more attention in the press. One consulting firm reported that 54 percent of Walmart's customers believed that Walmart acted "too aggressively," that 82 percent wanted the company to "act like a role model for other businesses," and, perhaps most damningly, that between 2 and 8 percent of customers had stopped shopping at Walmart because of "negative press they have heard." When I first suggested to my son that Walmart was one of the more sustainably focused firms that I knew, he looked at me skeptically. "I believe you, Mom," he said, "because you're my mom. But no one else will."

Reflecting on this storm of criticism later, Scott suggested that he was slow to recognize it because he believed that the negative feedback was coming from "blue-state elites," who didn't shop at Walmart and therefore didn't understand the money the company saved consumers. Andy Ruben, Walmart's first head of sustainability, recalled that inside Walmart, "You felt like you were in a bunker of some sorts and there was enemy fire every time you stuck your head up. The dissonance was so great between what I saw happening—people with such great intentions, what their aspiration was and what they were doing—from the way that that company was now being perceived outside of Bentonville."[36] (Walmart's HQ is located in Bentonville, Arkansas.) One expert who worked closely with the firm later recalled: "They were so isolated in Bentonville at that time they really didn't understand why people didn't just love them. They're like, 'We do everything right. We deliver to our customers every day the lowest price. We're hardworking. We have integrity.' That was their story."

In September 2004, Scott held a two-day off-site meeting focused on "the state of Walmart's world" and on how the firm might respond to its critics. At a subsequent meeting in December, the group agreed that it was time for Walmart to take a "strong stance on corporate responsibility." Eight months later Hurricane Katrina hit the Gulf Coast, giving Scott an opportunity to launch the new strategy in a particularly powerful way.

Hurricane Katrina was one of the most devastating disasters in US history. It flooded New Orleans and its surrounding communities, killed over one thousand people, created over one million refugees, and cost an estimated $135 billion.[37] Individual Walmart stores throughout the region—without waiting for orders from headquarters—began doing what they could for survivors, giving away food and clothing, and housing relief workers. One store manager in Waveland, Mississippi, "took a bulldozer and cleared a path into and through that store, and began finding every dry item she could give to neighbors who needed shoes, socks, food and water." On a conference call with his senior team, Scott told his team to respond without thinking of the quarter's budget, and Walmart corporate quickly got behind the local efforts, donating $20 million in cash, ten times the firm's initial pledge, along with one hundred truckloads of goods and one hundred thousand meals.[38]

The press praised Walmart's response to the storm at a time when government-led relief efforts in the early days of the crisis had largely failed. A *Washington Post* article entitled "Wal-Mart at Forefront of Hurricane Relief" noted, "The same sophisticated supply chain that has turned the company into a widely feared competitor is now viewed as exactly what the waterlogged Gulf Coast needs." Aaron Broussard, president of Jefferson Parish, Louisiana, appearing on *Meet the Press*, a nationally broadcasted Sunday morning television news show, stated, "If the American government would have responded like Wal-Mart has responded, we wouldn't be in this crisis."

The next month, in a speech broadcast to all Walmart's suppliers and to all its stores, offices, and distribution centers worldwide, Scott drew on Walmart's experience during the hurricane to announce a major commitment to sustainability. He introduced three key goals: to be supplied 100 percent by renewable energy, to create zero waste, and "to sell products that sustain our resources and environment," as well as a number of other commitments, including a 20 percent reduction in greenhouse gas emissions over seven years and a promise to double the efficiency of Walmart's transportation fleet. He also announced commitments to action with respect to health care, wages, communities, and diversity.

Scott set these goals in 2005. Sustainability was still a niche issue—something that only firms like Patagonia and Ben & Jerry's cared about. At the time Walmart's commitment was revolutionary. Recall that Unilever didn't announce its Sustainable Living Plan until 2010. It had immediate effects on public perceptions of the firm—as it was designed to. A 2008 report suggested that while in 2007 Walmart had ranked last in the ranking of twenty-seven retail companies by ethical reputation, in 2008 the company took third place (behind Marks & Spencer and Home Depot).[39]

But then something unexpected happened. Walmart found that saving energy was making the company a great deal of money. By 2017 Walmart had met its goal of doubling the transportation fleet's efficiency and was saving more than a billion dollars a year in transportation costs—around 4 percent of net income. Walmart doesn't release detailed investment figures, but in 2007 and 2009 we know that it was spending about $500 million on increasing energy efficiency and reducing GHG emissions. If it continued to spend at about this rate—and if the only benefits from this spending were the increased trucking efficiency—then my back-of-the-envelope calculations suggest that it received at least a 13 percent rate of return on its capital—at a time when many retail companies scramble to make 5 or 6 percent. Over the same period, Walmart has also increased the energy efficiency of its stores by 12 percent, which—again by my back-of-the-envelope conversion—is currently saving them about $250 million a year.

One could argue that this shouldn't have been a surprise. Engineers and consultants had been saying for years that there was money to be made in energy conservation. In 2007, for example, one of the world's leading consulting firms published a study claiming that the world could reduce its energy use by 25 percent if it simply adopted the energy saving measures that were profitable at the time.[40] Fine-tuning heating and cooling systems usually pays back within a year since about 30 to 40 percent of the energy used in heating and cooling older buildings is wasted.[41] KKR, one of the world's largest private equity firms, claims to have saved over $1.2 billion in energy costs and now routinely requires every firm that it buys to undergo an energy and water audit because the financial returns to such audits are so high.[42] There's now at least a billion dollar business in helping firms save money through energy reduction.[43]

But it took strategic vision to start uncovering these kinds of savings. This is not unusual. At Lipton, building a sustainable business model meant visualizing fundamental shifts in consumer behavior.

Walmart's breakthrough came from focusing on the everyday operational details of the business—from a profoundly different perspective. In its way, Walmart's commitment was just as transformative as Lipton's.

Playing the Odds in Energy

I want to come back to the concept of risk reduction as an economic case for pursuing shared value. We saw in the Unilever tea case that it can be important, but hard to quantify. When risks are common, and there is enough data to generate a good sense of their likelihood—in the case of house fires or automobile accidents for example—the costs of any particular risk can be quite precisely quantified. These are the calculations that built the insurance industry. But when risks are entirely novel, it becomes much harder to estimate how much one should pay to avoid them.

Perhaps this is why many firms have yet to integrate climate risk into their thinking, despite the fact that there's clearly something going on. Since the 1980s, for example, the scale of weather-related insurance losses has risen fivefold to about $55 billion a year. Uninsured losses are twice as much again.[44] One recent exercise suggested that the insurance industry may still be underestimating potential losses from extreme weather by as much as 50 percent.[45] Sea levels in Miami have risen by about six inches in the last thirty years. Current projections suggest that they will rise another six by 2035. Twelve inches of sea level rise in combination with spring tides or hurricanes is likely to cause catastrophic damage to oceanfront property.[46] Even with all this information readily available, I know that at least one major bank is underwriting mortgages on beachfront Florida real estate without including this possibility in its calculations of property values.

In April 2019 Mark Carney, the governor of the Bank of England, and François Villeroy de Galhua, the governor of the Banque de France, issued a joint statement pointing out that insured losses from extreme weather events have risen five-fold in the last thirty years. They suggested that the financial markets faced the risk of a climate "Minsky moment"—a reference to the work of the economist Hyman Minsky, whose analysis was used to show how banks overreached themselves before the 2008 financial crisis, and warned that those companies and industries that failed to adjust to climate change might cease to exist.[47]

In October 2019, Jerome Powell, the Federal Reserve chair, wrote to Senator Brian Schatz, noting that climate change was being "considered as an increasingly relevant issue for the central bank." The same month the San Francisco Fed published a collection of eighteen papers exploring the risks that climate change posed to the financial system.[48]

How can individual firms think about these kinds of risks, and build a case for action around them? One strategy is to think about an investment in sustainability not as a leap into the unknown, but as a strategic hedge. Take, for example, the case of CLP.

In 2004, CLP, one of the largest investor-owned utilities in Asia, announced that by 2010, 5 percent of its power would come from renewables. In 2007 it doubled down on this commitment by promising that 20 percent of its generating portfolio would be carbon free by 2020. These targets were the most ambitious of any power company in Asia—and by many conventional measures they made no sense at all.[49]

A majority of CLP's power stations were coal fired. This was not unusual in Asia, where coal is the fuel of choice for electricity generation since it's easily available and relatively cheap. In 2007 coal-fired power was significantly cheaper than power from solar, wind, or nuclear sources. In 2013—after significant declines in the cost of both solar and wind—CLP was assuming that wind would cost 30 percent more than coal, and that solar energy would cost three times as much.

What were they thinking?

I think that CLP was focused on the risks of remaining heavily dependent on coal-fired power. Sticking with coal presented significant political risk. Power plants are immovable, long-lived, and very expensive. They typically take three to five years to build and generate power for twenty-five to sixty years. Since they can't be moved and are often the only energy provider in an area, their success is critically dependent on maintaining good relationships with the local community, or on what is often called a "license to operate." CLP believed that there was a real possibility that at some stage local communities were going to start blaming local coal-fired power stations for the pollution in their neighborhoods and the flooding of their cities, and that this could put their license to operate in serious jeopardy. They feared that governments might move to penalize coal-fired stations—perhaps by increasing the price of coal through some kind of carbon price or tax, perhaps by simply shutting them down.

Sticking with coal also created technological risks. CLP believed there was a real possibility that solar and wind costs would drop precipitously. Most new technologies are expensive when they are first introduced. The first portable consumer cell phone, for example, cost consumers $3,995 in 1983 dollars (in 2018 money that would be just over $10,000).[50] But most technologies go down what's called a "learning curve." As demand increases, firms invest more in R&D, and as the technology is increasingly embedded in products, firms get better and better at making them. While in 2007 solar and wind power were both still much more expensive than coal, there was clearly some probability that sooner or later they might well be cheaper.

I don't know that CLP ever assigned a precise probability to either risk. But in 2008, when I asked a number of utility company executives what odds they would put on them, I got remarkably consistent estimates. Most of the executives believed that there was about a 30 percent chance that renewables would be cost-competitive with fossil fuels in the next twenty years, and about a 30 percent chance that public pressure would force governments to put some kind of tax or price on carbon. We mapped these two uncertainties into a 2x2 that looked like this:

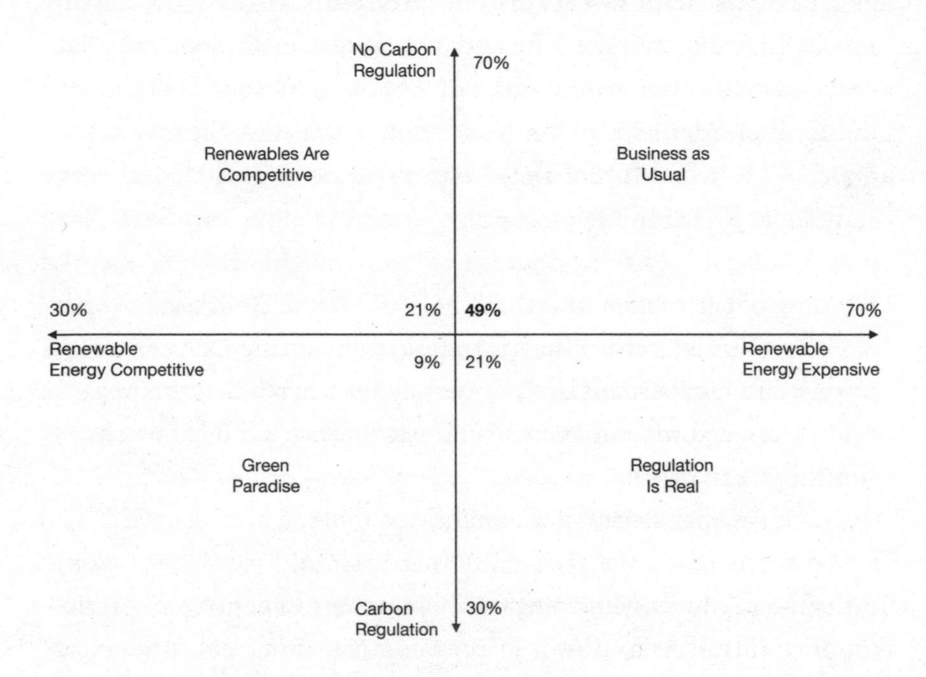

The top right-hand corner of this diagram defines "Business as Usual," a world in which there is no carbon regulation, and renewable energy remains expensive relative to fossil fuels. Sometimes I call this the "cross-fingered" future since so many firms spend so much time hoping that this is indeed where we're headed. According to the energy executives, there was about a 49 percent (.70 * .70) probability that in 2030 the world would look much like it does today. The lower left defines "Green Paradise"—a world in which carbon is priced and renewable energy is cheaper than coal. In 2008 many utility executives thought this was a very unlikely future, giving it only a 9 percent probability. But the odds of the other two futures—"Renewables Are Competitive with Fossil Fuels" and "Regulation Is Real" were about 21 percent each.

There were two interesting things about this diagram. The first was that it suggested that the odds of the future looking like the present were less than 50 percent. The second was that this was almost always news to everyone in the room. Usually the group moved from ridiculing the visionaries who were convinced that Green Paradise was imminent, to wondering how to hedge their bets. CLP's leadership thought the odds that the world would see a business-as usual future were very low indeed. In 2013 Andrew Brandler, the CEO of CLP phrased it this way:

> We see carbon as a long-term threat to any business. In 2050, if you are a carbon-intensive business, you are in big trouble; chances are you won't be in business by then. We have been in business for over 100 years, and we want to be in business in 2050, but that doesn't mean you take action in 2049. You have to move down this path and be ahead of the curve as the world moves.

This is the key to understanding CLP's strategy. The flip side of risk is opportunity. If Asia's power sector was going to decarbonize—and CLP believed that it was—moving to carbon-free energy ahead of the competition was potentially an exceedingly attractive business opportunity. Fifteen years later, their early commitments look prescient. Between 2010 and 2018, for example, the global weighted-average cost of solar- and wind-generated electricity has fallen by 35 and 77 percent, respectively.[51] Installation costs have dropped by 22 and 90 percent.[52]

In some places solar and wind are already cheaper than coal. They are "intermittent" sources of power—they only work when the sun is shining and the wind is blowing—so using them to replace fossil fuels at scale requires further reductions in the cost of storage. But the rate at which their costs are falling is quite breathtaking, and those of my colleagues who work in renewable energy tell me that the odds that renewable energy will be cost competitive with coal by 2030 are very high indeed.[53]

Renewables made up 38 percent of China's total installed generation capacity by the end of 2018,[54] and current projections are for China's electricity grid to be 60 percent renewable by 2030,[55] and for India's to be 67 percent renewable by 2050.[56] $6 trillion of new energy investment will go into Asia Pacific in the next twenty years—new capacity in China will be greater than that in the United States and Europe combined, and new capacity in India will be greater than either the United States or Europe.[57] Carbon-free power is a huge market opportunity, and thanks to its early investments in the field, CLP is particularly well positioned to succeed. For CLP, tackling risk has meant huge opportunity.

If It's So Great, Why Isn't Everybody Doing It?

Between them Lipton, Walmart, and CLP put the economic case for creating shared value clearly on the table. Reducing environmental damage and treating people well reduces reputational risk. It assures the long-term viability of the supply chain. It can persuade consumers to favor your products and services over those of your competition. It can reduce costs. It can create entirely new businesses—particularly if, like CLP, you are sophisticated enough to see how the world is changing before others do.

Robin Chase founded Zipcar—a car sharing service—in 2000, nearly twenty years ago, years before the rest of us discovered the sharing economy. She saw Zipcar as part of a much larger vision for how the economy might be transformed. In one interview she explained:

> The collaborative economy is larger than the sharing economy. The sharing economy feels to me like it's about assets. The collaborative

> economy is everything. It's making clear and visceral to us that, if I can have real-time access not just to hard assets, but to people, to networks, to experiences, it means that the way I do my own personal life is completely transformed. I don't have to do any hoarding.
>
> I don't have to be worried about having stuff and owning it. I can start to rely on the fact that I can reach out and find the right person at the right moment. That dramatically transforms how you live. Instead of on-demand cars, it's an on-demand life, in a much larger fullness.[58]

Zipcar grew to be the largest car sharing company in the world and was acquired by Avis in 2013 for half a billion dollars.[59] It now has over a million members in five hundred cities in nine countries. Since leaving Zipcar, Chase has founded or helped to found at least three other ventures with similar goals—Buzzcar, a peer-to-peer car sharing service; GoLoco.org, a ride sharing company; and Veniam, a firm that uses cars and trucks to blanket a city with public Wi-Fi.

But every time I teach, people ask me if firms can really make money by doing the right thing. "I know about Tesla," they say, "but are there any other examples out there?" I tell them that there are hundreds and refer them to the Harvard Business School case website, armed with the right key words. But it is certainly the case that shared value has yet to go entirely mainstream. Why? Why are so many businesspeople so reluctant to think that worrying about how they treat their people or the environment around them might be a powerful source of profitability?

The key to this puzzle, I believe, is to realize that the embrace of shared value is, first and foremost, an innovation—and more precisely, an *architectural* innovation. Architectural innovations change the relationship between the components of a system—the system's architecture—without changing the components themselves. And because most people in most organizations are focused on the components of the system they're embedded in, rather than the relationship between them, architectural innovations are hard to spot and hard to react to. Architectural knowledge—knowledge of how the components fit together—becomes embedded in the structure, in the incentives, and in the information processing capability of the organization, where it becomes effectively invisible, making it very difficult to change.

Much of the conversation around innovation focuses on the potential for cool new technologies to disrupt existing industries. We tend to think of the way artificial intelligence is likely to change the world, or of the potential for algae grown in tanks to replace oil. But Michiel, Scott, Andrew, and Robin are also architectural pioneers—inventors of new ways of thinking about the structure and purpose of the firm. The idea that the way to grow Lipton's tea business was to increase the price it paid for tea was a profoundly revolutionary one, requiring a quite new way of thinking about the entire value chain. Walmart was famous for its skills in cost reduction. Who could have imagined that thinking about something as nebulous as saving the environment would lead to previously undiscovered billion dollar cost-reduction opportunities? CLP committed to transforming their entire business—at a time when the available alternatives cost significantly more and came festooned with as yet unsolved technical problems. Robin invented a new business from whole cloth because she saw the purpose of the economy in an entirely different way. Creating shared value is an act of profound imagination. If you're deeply embedded in old ways of doing things, it's hard to grasp the benefits of something as bold as reimagining capitalism.

Take, for example, the case of Phil Knight, the founder of Nike, one of the most successful entrepreneurs of the last fifty years. Phil completely revolutionized the footwear and apparel businesses. But he steadfastly ignored the risk that child labor in his supply chain presented to Nike's business, and in doing so put Nike's brand—its most valuable asset—at considerable risk. He had a strong business case for embracing shared value, and he missed it completely.

What was he thinking?

Phil drove Nike's growth through three core insights. The first was that the way to drive down costs was to subcontract production to cheaper locations overseas. In the 1970s this was a revolutionary idea. The second was that continuous innovation was the key to success. Nike invested heavily in research right from the beginning. The third was the one that supercharged Nike's success: the power of marketing. Phil—who understood both the symbolic power and attractiveness of sports years before almost anyone else—channeled the bulk of the money saved by Nike's production strategy into its marketing budget. In the words of one journalist:

> Nike is a cultural icon because Phil understood and captured the zeitgeist of American pop culture and married it to sports. He found a way to harness society's worship of heroes, obsession with status symbols and predilection for singular, often rebellious figures. Nike's seductive marketing focuses squarely on a charismatic athlete or image, rarely even mentioning or showing the shoes. The Nike swoosh is so ubiquitous that the name Nike is often omitted altogether.[60]

In combination the three ideas turned out to be commercial dynamite, and by 1992 Nike had $3.4 billion in sales. But Phil was not an entirely happy man. His investors seemed to be oblivious to the power of his vision, no matter how often he tried to explain it to them.

In Nike's Annual Report for the year, for example, he noted that Nike had $3.4 billion in sales and was the largest athletic shoe company in the world. That it "had made every major advance in athletic shoes in the last twenty years" and "broke[n] the billion dollar market in international sales for the first time in a single fiscal year." And yet "for all but a couple of very brief periods, we have always sold at a substantial discount from the S&P's 500 Index P/E multiple. . . . We get summed up with the old label 'sneaker company' . . . [and] lumped into the apparel category."[61]

A company's P/E multiple is the ratio between its market value and its after-tax earnings. In general, investors give companies they believe are likely to grow substantially higher multiples. In the late 1990s and early 2000s, for example, health care, IT, and telecoms had much higher P/E multiples than the rest of the economy. To put Phil's annoyance in perspective, Amazon's P/E has never been below fifty-six, and is now over a hundred. But Nike's P/E didn't reliably break twenty until 2010.[62] In short, Phil believed that his investors had no idea how quickly the company was going to grow, and were underpricing it as a result.[63] For the next five years the idea continued to be central to his annual letter.

In 1993 he wrote: "Nike continues to be an undervalued global power brand. . . . Athletic shoes and clothes—especially shoes—are not commodities. Try this: run a marathon, or even a mile, in a pair of $19.95 Wal-Mart specials. That will end the discussion."

In 1994: "Although it was our first down year in seven years we generated an 18 percent ROI. You'd think that for a company for which an 18 percent return was labeled 'bad' would get higher than a 15 P/E multiple when the market average stood at 20, wouldn't you?"

In 1995, after noting that this is "the greatest year in industry history," he wrote: "Even in good times such as these, as I sit down to write this letter I become angry and frustrated. . . . If you show this record to investment analysts without identifying the company, an experiment we have done, they will say it deserves a P/E multiple in excess of the S&P 500. Only after the company's name is revealed does the consensus move to a multiple at a discount from the market. . . . This has reached the point of ridicularity."

In 1996 Nike's multiple rose, but Phil was still not entirely happy: "By (the) standards of measurement we normally use in this space, fiscal year 1996 was a fantastic year. We set an all-time record for sales and earnings. . . . And to its credit, Wall Street saw fit to increase our multiple. . . . The central question was: does a fashion company, no matter how strong the brand, deserve such a multiple? The problem with the debate is that the answer doesn't matter. It's the question that is wrong."

Throughout the early 1990s, in short, Phil grappled with a problem that besets many successful, visionary entrepreneurs. He couldn't seem to communicate the power of his vision to his investors. Knowing what we know now, of course, reading his letters makes clear just how prescient Phil was.[64] He talks extensively about the ways in which sports will create worldwide brands. He explains—again and again—how he's investing for a global future—how Nike's investments in innovation, in endorsements, and in building infrastructure on the ground overseas are going to yield enormous dividends. But until 1996—when even the slowest analyst began to wake up to the Nike phenomenon—he couldn't get the multiple up beyond the S&P average. It's no wonder he got a little frustrated.

Yet at the same time that Phil was raging against the blindness of Wall Street analysts, he himself was proving to be equally blind to something that was about to shake his business to its roots.

In 1992 *Harper's Magazine* published the paycheck of Sadisah, a young Indonesian woman who worked for the Sung Hwa Corporation, making shoes for Nike. The piece—which was written by Jeffrey Ballinger, a worker's rights advocate who had spent nearly four years in Indonesia—showed that she was paid about $1.03 a day, or just under 14¢/hour. Ballinger suggested that at these rates the labor cost embedded in an $80 pair of shoes was approximately 12¢, and closed his piece by asking:

> Boosters of the global economy and "free markets" claim that creating employment around the world promotes free trade between industrializing and developing countries. But how many Western

> products can people in Indonesia buy when they can't earn enough to eat? The answer can't be found in Nike's TV ads showing Michael Jordan sailing above the earth for his reported multiyear endorsement fee of $20 million—an amount, incidentally, that at the pay rate shown here would take Sadisah 44,492 years to earn.[65]

In 1993 CBS aired a report detailing abusive working conditions at Nike's Indonesian suppliers, and in 1994 a series of harshly critical articles appeared in *Rolling Stone*, the *New York Times*, *Foreign Affairs*, and the *Economist*. In 1996 *Life* magazine published a devastating exposé of child labor practices in Pakistan and India. The author described going undercover as an American interested in setting up shop in Pakistan to make soccer balls for export. He found children working as slaves, unable to leave because they could not afford to pay off the *peshgi* their parents had taken for them. One foreman offered to get him "as many as 100 stitchers if you need them," noting that "of course you'll have to pay off their peshgi to claim them." Children were blinded, malnourished, beaten if they asked for their parents and—most evidently—hardly paid at all. Schanberg, the author of the article, claimed that the average child laborer made 60¢ a day. At the head of the article ran a photograph of a twelve-year-old boy stitching Nike's soccer balls. The implication was clear: Nike was employing enslaved children.

Doonesbury, the popular comic strip, devoted a full week to Nike's labor issues. Student organizations at many campuses urged boycotts of Nike's products. Nike was in the midst of expanding its chain of giant retail stores, and found that "each newly opened NikeTown came with an instant protest rally, complete with shouting spectators, sign waving picketers and police barricades." Nike's celebrity endorsers—including Michael Jordan and Jerry Rice—were publicly hounded.

But throughout the fracas, Nike continued to insist that what happened in its supply chain was none of its concern—that it had a code of conduct in place that prohibited abusive behavior, and that its suppliers were independent contractors over whom it had no control. Neal Lauridsen, Nike's vice president for Asia said, "We don't know the first thing about manufacturing. We are marketers and designers."[66] John Woodman, Nike's general manager in Jakarta, explained, "They are our subcontractors. It's not within our scope to investigate [allegations of labor violations]." Adding, "We've come in here and given jobs to thousands of people who wouldn't be working otherwise."[67]

Labor issues don't show up in Phil's letter to shareholders until 1994, when he took particular exception to a *New York Times* piece by the sports writer George Vecsey, complaining that it was "two columns of incessant railing on Nike, the terrible entity." In 1995 there was no mention of the supply chain. In 1996 he said the following:

> By the standards of measurement we normally use in this space, fiscal 1996 was a fantastic year. . . . Yet no sooner had the great year ended than we were hit by a series of blasts from the media about our practices overseas. So I sat with a dilemma: Use this space to answer our critics' misconceptions, . . . or try to give our owners the bigger picture of their company. . . . I chose the latter.[68]

In 1997 he was captured on camera talking to the director Michael Moore in a documentary called *The Big One*:

> **Moore**: Twelve-year-olds working in [Indonesian] factories? That's O.K. with you?
>
> **Knight**: They're not twelve-year-olds working in factories . . . the minimum age is fourteen.
>
> **Moore**: How about fourteen then? Does that bother you?
>
> **Knight**: No.

There is no word about labor issues in the supply chain in his annual letter.

Then profits collapsed. Nike had been growing very fast. In 1997 revenues were up 42 percent and net income by 44 percent—but in 1998 demand cooled. Nike's critics suggested that public outrage over its labor practices was partly to blame. In 1997 there had been just under three hundred pieces published coupling the words "Nike" with "sweatshop" or "exploitation" or "child labor."[69]

In May 1998, at a speech at the National Press Club, Phil changed his tune, acknowledging that "The Nike product has become synonymous with slave wages, forced overtime and arbitrary abuse."[70] He announced the formation of a corporate responsibility function at Nike, and committed the company to a number of new initiatives designed to improve factory working conditions, including raising the minimum wage; using independent monitors; strengthening environment, health, and safety regulations; and funding independent research into conditions in Nike's supply chain. Nike is now one of the leaders of the push to make apparel supply chains more sustainable, and independent rankings routinely rank Nike as the most sustainable footwear and apparel company in the world.[71]

Nike's story is the key to understanding why so many firms are having trouble making the switch to shared value. It took Phil, a visionary entrepreneur who could see things almost no one else could see, five years to understand the threat that problems in the supply chain posed to his brand—even as he was berating his investors for not understanding the roots of Nike's success. The irony here is that in failing to understand that the world was changing in ways that made it essential for Nike to pay attention to labor conditions in its supply chain, Phil was falling into precisely the same kind of error as his investors—who were failing to understand the ways that sports were remaking the footwear and apparel businesses.

When the world shifts in unexpected ways, even the most visionary businesspeople have trouble understanding what's happening.

Phil's investors missed Nike's potential, and Phil and his colleagues missed the supply chain issue because they were both architectural innovations—changes in the way the pieces of the puzzle are put together.

Incremental innovation—innovation that improves one particular piece of the puzzle—is often harder to deliver than it sounds, but it's usually easy to see that it must be done, and it doesn't threaten the status quo. Nike under Phil was an expert incremental innovator, introducing significantly improved running shoes year after year.

"Radical" or "disruptive" innovation tends to get most of the attention. These are innovations that make old ways of doing things completely obsolete. Think of digital photography or of the new drugs that stimulate the patient's own immune system to fight cancer. But while radical innovation presents a profound challenge to successful organizations, it's a challenge whose scope is immediately clear. Digital photography forced Kodak into bankruptcy, but not because Kodak failed to see the threat that it represented. Indeed, Kodak invested deeply in digital photography right from the beginning and made a number of groundbreaking discoveries in the area.

It's architectural innovation that creates trouble, and it's architectural innovation that brought Kodak down. The move to digital photography changed the architecture of the product—cameras became a component of telephones instead of stand-alone machines that had to be lugged around—and the ways in which photographs were shared, printed, and used. Kodak found it all but impossible to adapt. Radical innovation is tough but visible. All the major pharmaceutical companies can see that understanding genetics is going to be central to finding new drugs, and all of them have invested heavily in bringing genetics into their research operations. But architectural innovation runs under the radar. It often looks as if it's about a relatively small change to one small piece of the puzzle, but it's actually about a complete rethinking of the way the pieces fit together.

Tim Harford, the *Financial Times*' "Undercover Economist," uncovered a wonderful example of the ways in which successful organizations can miss the power of architectural innovation in the history of the British response to the invention of the tank.[72] The tank was invented by E. L. de Mole, an Australian who approached the British War Office with his design in 1912, two years before World War I broke out.

By 1918, the last year of the war, Britain had the best tanks in the world, and the Germans none at all. Indeed the allies forbade their production. But by the 1930s, Germany had leaped ahead, and in 1939, the first year of the Second World War, it produced twice as many tanks as the British—and used them to much greater effect.

The problem was classically architectural. The British Army didn't know where to put the tank. It was divided into two great branches—cavalry and infantry. The cavalry's job was to be swift and mobile. Tanks were swift and mobile. Perhaps the tank was a special kind of horse and belonged with the cavalry? The infantry's job was to be an immovable source of invincible firepower. Tanks were tough to dislodge and extraordinarily powerful. Perhaps the tank was just a very strong infantryman with a particularly powerful gun? One could, of course, start an entirely new unit, just for tanks. But who would fight for such a unit? Who would fund it?

Of course the tank was neither simply a faster kind of horse nor a more powerful kind of infantryman. It was some kind of cross between the two—only more so—with the potential to enable a completely different kind of warfare. A British army officer named J. F. C. Fuller recognized this potential while the First World War was still underway. In 1917 he presented a detailed plan to his superiors, suggesting that tanks coupled with air support could do an end run around the German trenches, attacking German headquarters behind the line and ending the war almost immediately.

Fuller's biographer calls his idea "the most famous unused plan in military history," but of course it was used—in 1940, by the Germans, who called it "blitzkrieg."

The British had given control over the tanks to the cavalry, and the cavalry was orientated toward its horses, rather than toward the new weapon. The horse was the central element of the cavalryman's life—his pride, his joy, his reason for being. Field Marshall Sir Archibald Montgomery-Massingberd, the most senior general in the British Army, responded to the threat of Nazi militarization by providing each cavalry officer with a second horse and increasing spending on horse forage tenfold. The United Kingdom entered the Second World War seriously unprepared to respond to a competitor who had redesigned its army around the tank, rather than the other way around.

Architectural innovation is difficult to see—and often hugely difficult to respond to—because in nearly every organization, most people spend the vast majority of their time paying attention to the piece of the puzzle they have been assigned. If you're a door handle engineer at a major car company, you pass your days designing door handles. You go to door handle conferences and follow door handle trends. You don't spend a lot of time thinking about how the automobile industry as a whole is likely to change. To survive, everyone develops mental models of how the world works that tell us what we need to pay attention to and what we can safely ignore.

You'd think that CEOs—who are supposedly paid to think about the big picture—wouldn't fall into this trap. But as Nike's case illustrates, they do. Indeed, in Phil Knight's case, one way to think about what happened is that it was precisely because he was putting so much time and energy into communicating his vision of the future (and of course, into building a multibillion dollar global firm) that he didn't have the mental space to appreciate the significance of what was happening in his supply chain. Phil and his colleagues' complete conviction that they had no responsibility for things that happened beyond the boundaries of the firm was so deeply rooted that they found the initial criticism they faced almost incomprehensible. They "knew" that their responsibility to their employees stopped at the boundaries of the firm. It was just a given—the way the world operated—an assumption shared by nearly every businessperson they knew. In a way that now seems deeply ironic, at the very moment that Phil was berating his investors for failing to understand that the deepest assumptions underlying the sports business were changing in profoundly important ways, he himself couldn't see that old assumptions about the nature of the relationship between a brand and its supply chain were also changing, and that child labor in his supply chain posed a massive threat to his brand. Once the world sent him a sufficiently strong signal that the issue could no longer be ignored, he responded—as we would expect—with energy and skill.

There is enormous opportunity to create shared value. Individual firms can address environmental and social problems and build thriving businesses at the same time, by reducing their costs, protecting their brands, ensuring the long-term viability of their supply chains, increasing demand for their products, and creating entirely new businesses.

But these opportunities can be hard to see. Building a just, sustainable society is going to be as disruptive as moving from steam to electricity, or learning to use the internet or to take advantage of AI. Firms that are doing well under the current dispensation will claim that there's no need to change—that if there is a need to change, there is no business case—and that even if there is a business case they're too busy to work on it right now. This is what change looks like.

When I was at MIT, I held the Eastman Kodak Professor chair, and spent some time working with Kodak as it attempted to respond to the threat of digital photography. It had no trouble making the technological transition. A Kodak engineer was the inventor of the first digital camera, and the firm held many of the early patents in digital photography and built a large digital camera business. But Kodak was unable to develop a business model that would allow it to make money: consumers printed digital photographs far less often, and the firm did not anticipate the way in which the camera would become an integral part of the mobile phone. Kodak went bankrupt in 2012, the victim of profound architectural innovation.

I've spent more than twenty years of my life studying these kinds of change. I've learned at least three things. The first is that recognizing and responding to architectural innovation is hard but not impossible. Phil Knight had trouble—perhaps precisely because he was so successful—but Lipton, Walmart, and CLP were all able to use the creation of shared value as a route to significant competitive advantage. The second is that those firms that manage to take advantage of these kinds of transitions—that have the courage to invest before their competition and to invest in the skills and people required to build entirely different ways of approaching a market—have the potential to reap enormous returns.

The third is that organizational purpose is the key to change. Those firms that have a clearly defined purpose beyond profit maximization, where it is clearly understood that the purpose of the firm is not to make shareholders rich, but to build great products in the service of the social good—these are the firms that have the courage and the skills to navigate transformation.

Redefining the purpose of the firm is central to reimagining capitalism. What exactly this means and what it might look like in practice is the subject of the next chapter.

4

DEEPLY ROOTED COMMON VALUES

Revolutionizing the Purpose of the Firm

> Some people think greed is good. But over and over it's proven that ultimately generosity is better.
>
> —PAUL POLMAN, RETIRED CEO OF UNILEVER[1]

On January 12, 2015, in a packed hotel ballroom in Jacksonville, Florida, Mark Bertolini, the CEO of Aetna, announced that beginning in April the firm would be paying a minimum wage of $16/hour.[2] Aetna was one of the largest health insurance companies in the world, and his move made headlines. Nearly six thousand employees—around 12 percent of Aetna's domestic workforce—would see their pay increase by an average of 11 percent. Some would receive a pay raise of 33 percent.[3] Mark also announced that many of these employees would be able to sign up for Aetna's richest health benefit plan for the price of the cheapest. As a result, some employees would see an increase in their disposable income of more than 45 percent. The ballroom exploded. Mark later said: "I had known people would be happy, but I wasn't ready for the raw emotion. There were people crying. People saying, 'Praise the Lord. My prayers have been answered.' The frontline managers were thrilled."

It was an expensive move, increasing Aetna's labor costs by about $20 million a year, and it did not go down well with some of his colleagues in senior management. Among Aetna's minimum-wage employees, 80 percent were women. Most of them were single mothers, and some were on food stamps and/or Medicaid. But when Mark proposed paying them $16/hour, he faced stiff resistance, much of it framed in classic shareholder value maximization terms. In his words, "They told me we'd be paying above-market rates, particularly in states with lower-than-average wages. We'd be hurting our bottom line. We have shareholders to serve. We have Wall Street to satisfy."

What was Mark thinking?

One interpretation is that he was simply indulging his own sense of morality at Aetna's expense. When asked, Mark did indeed frame the wage increase as a personal, deeply moral decision. He talked about the ways in which his decision to become active on social media had alerted him to the struggles many of his employees were facing. "More and more often, I saw people online saying, 'I can't afford my benefits. My healthcare coverage is too expensive.'" He told the *New Yorker* that it was not fair for employees of a Fortune 50 company to be struggling to make ends meet, and explicitly linked the decision to the broader debate about inequality, mentioning that he had given copies of Thomas Piketty's *Capital in the Twenty-First Century* to all his top executives. "Companies are not just money-making machines," he told the magazine. "For the good of the social order, these are the kinds of investments we should be willing to make. There definitely is a moral component and, you know, I had plenty of arguments that the spreadsheet wouldn't pencil out. And my view was, in the end, this is just not fair."[4]

But this was only part of the story. Behind the scenes, for both personal and professional reasons I explore below, Mark was in the midst of an audacious strategy to create shared value by fundamentally transforming Aetna's business model. Paying each of his employees a living wage was a key element of this strategy—a move to create shared purpose and in doing so, create the commitment, creativity, and trust that would enable him to implement his vision.

The widespread adoption of authentic purpose—a clear, collective sense of a company's goals that reaches beyond simply making money and is rooted in deeply held common values and embedded in the firm's strategy and organization—is an essential step toward reimagining capitalism. It has three critically important effects. First, deeply embedded authentic purpose makes it much easier to identify the kinds of architectural innovations that enable the creation of shared value. Second, it makes it much easier to take the risks and find the courage required to actually implement these kinds of innovation. Third, building a genuinely purpose-driven organization is in itself an act that creates shared value, since it requires creating the kinds of jobs that are needed to begin to address inequality and build a just society.

Mark's business case began with the fact that the US health care system is in trouble. US health care costs nearly twice as much (measured as a percentage of GDP) as health care in the rest of the developed world without delivering notably better results.[5] In one study, for example, the World Health Organization placed the United States thirty-seventh out of 191 counties in terms of its "overall health system performance." Another review evaluating the health care systems of eleven countries—Australia, Canada, France, Germany, the Netherlands, New Zealand, Norway, Sweden, Switzerland, the United Kingdom, and the United States—suggested that the United States had the worst performance of the group.[6]

At the same time Aetna's existing business was coming under increasing stress. Health insurance is one of the least well-regarded industries in the United States, with a net promoter score below that of airlines and cable TV companies.[7] Moreover, in the face of increasing economies of scale, it was steadily consolidating—and Aetna was a distant third to the two industry leaders, United Health and Anthem.[8] Mark needed a new strategy, and his own deep sense of purpose guided his choice.

Mark's sense of mission was triggered by two life-shattering events that occurred in his forties. In 2001 his sixteen-year-old son Eric was diagnosed with terminal cancer. He later said, "What I was told was that he had six months and no one had ever survived his cancer."[9] Mark quit his job, and according to one observer, "all but moved into his son's hospital room, torturing the medical team for information and helping his son get an unapproved drug." Another noted that Mark "downloaded a copy of *Harrison's Principles of Internal Medicine*, a bible for junior doctors, and started having fierce arguments with the medics, who thought he was in denial about his son's chances of survival." At one moment Eric nearly starved to death because he was allergic to the only fat supplements approved for use in the United States, but Mark persuaded a doctor to locate a fish-based supplement in Austria, filed for the regulatory exemption, and persuaded the maker's chairman to bring it to the United States on his next flight there. His son is the only person to have ever survived his type of cancer—gamma-delta T-cell lymphoma.[10]

The episode shaped his views of the US medical system and of what was needed to fix it. "Lesson one was that they always viewed him as the lymphoma in room four, whereas I knew him from the delivery room when he was born. . . . Their view of him was as a disease, not a person. . . . What I learned through that experience was that the health care system is not very connected," said Mark. "We were the connection. We were the advocates."

This perception that the medical system cared about individual procedures and the bottom line rather than about its patients as whole human beings was reinforced when in 2004—less than a year after he had joined Aetna—he suffered a life-threatening skiing accident that broke his neck in five places and left him with permanent arm damage. The medical establishment prescribed painkillers, but in his words:

> During the recovery, I'm on seven different narcotics all at once. Fentanyl patches, Vicodin, OxyContin, Neurontin, Keppra. And liberal use of alcohol when I didn't have to go anywhere. It was a mess. Somebody suggested Craniosacral therapy. I said, "What the hell is it?" But by the fourth visit I was feeling better, and over a period of five, six months got off all of my drugs. I got hooked on Craniosacral therapy. Then the Craniosacral therapist said to me, "You should try yoga." I said, "Ah, that's for girls." But after I tried it, I couldn't move the next day. I said, "Oh my god. This is amazing, what a workout." I started practicing every day because it made me feel better. And about two months into it I said, there's more to this. So I started reading the Upanishads, the Bhagavad Gita, went to retreats, learned some chanting, studied some Sanskrit, and was like, "This is like amazing."[11]

In response, he began to drastically reconfigure Aetna's strategy. He hoped to use Aetna to transform his members' health care by making it much more personal and much more connected. He set up two distinct initiatives. The first was the creation of a leading-edge digital platform built on big data and world-class behavioral economics. The platform would not only simplify the way in which Aetna's members interacted with Aetna (a major pain point in the current business) but also offer a range of applications that would support Aetna's members in taking care of their own health in real time. In the United States, for example, 20 to 30 percent of medication prescriptions are never filled and approximately half of all the medications prescribed for chronic disease are never taken.[12] This leads to approximately 125,000 deaths a year[13] and increases health care costs by between $100 billion and $289 billion annually.[14] One senior member of Mark's team described the way in which the platform might help fix this problem:

> A simple thing that we can do is institute a reminder program, and specifically target members during the first six prescription fills. We want to give them incentives to adhere to their medication, and we can run a vast number of experiments to rapidly test ideas and learn what's going to move the needle. Do we give a member in this reminder program the incentive up front, or do we wait until they've filled their sixth prescription? Ultimately, we want to be able to offer the right incentive to the right person at just the right time, and in a way that they want. For example—and this is dependent on the level of consent that the member has given us, and who we might partner with—if a member is walking past a CVS or some other retail partner, we could send them a message on their Apple Watch to go get a flu shot, and if they do, they'll get some form of reward.[15]

The second initiative was to put people on the ground who could work face-to-face with Aetna's sickest members. In the first set of pilots, for example, the firm put a multidisciplinary team into each of eight districts in Florida. Each team included nurses, pharmacists, behavioral health experts, social workers, dieticians, and community health educators. All Aetna members were assigned a field care manager, whose job was to reach out to them, to learn about their goals for their health, and to bring in other parts of the team as needed to reach these goals. Christopher Ciano, president of Aetna's Florida operations, explained:

> The Aetna Community Care program uses a holistic approach to truly understand the needs and goals of each individual member. A comprehensive and personalized plan is then designed to address those needs. Historically, many of our programs have been

designed around disease states, and not [around] member specific personalized goals. Take congestive heart failure for example. We previously weren't setting outcomes based upon the personal desires of the member—maybe our member just wants to be able to go outside and play with their grandchild instead of achieving some common disease state metric. Our new approach focuses on what each member wants in terms of their specific health ambitions, and we set out to help members achieve these goals by being in the community and interacting with them where they live, work, and play, rather than simply engaging with them telephonically or through the mail.[16]

Mark described the new strategy as "the consumer part of the health care revolution." At its heart was a classic shared value thesis: the belief that if Aetna could partner with its members to improve their health, not only would its members be much healthier, but Aetna's costs would fall, and Aetna would build a thriving, profitable—and highly differentiated—business. In the words of Gary Loveman, the man Mark hired to execute this strategy:

The general understanding that everyone has of health care is that it's hopelessly complicated and nearly unaddressable, and that tends to stop a lot of people in their tracks. I have a simpler idea, which is that many Americans are unnecessarily sick, and their costs are needlessly high and their lives needlessly challenged as a result. For example, take two 60-year-old men who have diabetes and early-stage renal failure. One of them follows medical counsel tightly and lives a productive, happy life with only slightly above-average healthcare costs. The other does not follow medical counsel and lives a very precarious, unhealthy, and costly life with a lot of trips to the hospital and the emergency room. My ambition is to get that second guy to look like the first guy. If I can do that, I can get a lot of people healthier, and we can all save an awful lot of money.[17]

Framed this way, of course, Mark's strategy looks simply like good business. But that's the nature of shared value—it's all about addressing the big problems while simultaneously building a business case. It's not a question of purpose *or* profit. It's about using the broader view that a larger purpose provides to find these kinds of opportunities—and then embedding purpose into the organization in a way that enables the firm to execute them.

Mark's strategy was risky. It required an audacious level of architectural innovation—a complete rethinking of how Aetna worked with its customers and how it would create value. You will recall from the story of the British Army's reaction to the tank just how hard architectural innovation can be to execute—particularly within a large, well-established, reasonably successful organization. For more than a hundred years, Aetna's business had been about selling and administering insurance. The firm had made money by controlling costs, not by advocating for its patients. Mark's strategy required everyone—from the senior leadership team to the people answering the phones—to develop a significantly different set of skills and to act in very different ways.

My experience is that the most reliable enabler of this kind of change is deeply held shared purpose. It aligns everyone in the organization around a common mission. It gives everyone a reason to work toward the goals of the organization as a whole, rather than their own personal goals. Most importantly, it unleashes the kind of creativity, trust, and sheer excitement that enables old firms to do new things.

People will work hard for money, status, and power—"extrinsic" motivators. But for many people, once their core needs are met, the sheer interest and joy of the work itself—"intrinsic" motivation—is much more powerful. Shared purpose creates a sense that one's work has *meaning*—one of the core drivers of intrinsic motivation and a driver of higher-quality, more creative work. It also creates a strong sense of identity, another source of intrinsic motivation and a powerful source of trust within the firm. To the degree that purpose supports authenticity—the ability to live a life in accordance with one's deepest values—it also increases the presence of positive emotions—something that is strongly correlated with the ability to see new connections, to build new skills, to bounce back after difficult times, and to be more resistant to challenges or threats. The employees of purpose-driven firms are thus likely to be significantly more productive, happier, and more creative than those at more conventional ones.[18]

An authentic purpose also turbocharges the ability to work in teams. Employees who are deeply identified with the firm's purpose share a common set of goals. They are also likely to be significantly more "pro-social"—that is, to be temperamentally inclined to trust others and to enjoy working with them. Teams that share common goals and that are composed of individuals who are truly authentic, fundamentally prosocial, and intrinsically motivated find it easier to communicate and align their activities, to trust each other, and to create a sense of "psychological safety"—all attributes that drive high performance, and the ability to take risks and to learn from each other. Purpose-driven firms are thus likely to be much more open to new possibilities and far more capable of handling the architectural change that is often required to take advantage of those opportunities.

Mark's own strong sense of purpose gave him the perspective and the passion necessary to design Aetna's new strategy. But unleashing the creativity, trust, and commitment required for Aetna to execute it required building a purpose-driven organization—one in which the vast majority of Aetna's employees were themselves deeply committed to the new purpose and convinced that Aetna's senior team were authentically driven by this purpose themselves.

Mark threw himself into this task. He began by communicating his personal story as often and as authentically as he could. He papered the walls of Aetna's headquarters with cheerful posters laying out the firm's new values. But talk is cheap. Persuading thousands of people that you're for real—that your foremost goal is making a difference in the world, rather than making a difference to the bottom line—requires making it clear that there are times when you will do the right thing just because it is the right thing. That you will—at least occasionally—put purpose before profit.

This is, I believe, how we should understand Mark's decision to raise Aetna's minimum wage. Arguing that Aetna's employees would be much more likely to become committed to member health if their own health were taken care of, Mark had introduced yoga and meditation classes at Aetna. Eventually every Aetna office with more than two thousand employees had an acute care center, a fitness center, a mindfulness center, and a pharmacy. He faced some pushback. He recollected,

> I had people pushing against it, with our C.F.O. at the time saying, "We're a profit-making entity. This isn't about compassion and collaboration." I said, "Well, I actually think it is. And I'm in charge, so we're going to do it."[19]

Then he raised the minimum wage. He was careful to talk about the economic case for doing so, noting that many of the affected employees worked in customer service, and arguing that more engaged employees would make better connections with Aetna's customers. As he said, "It's hard for people to be fully engaged with customers when they're worrying about how to put food on the table."[20] But that's not why he did it—or not only why he did it. He did it because he absolutely believed that it was the right thing to do. And here's the paradox. His willingness to take the heat that came with the decision was a telling signal of his authenticity—and that in turn, was an important step toward unleashing the power of purpose across the whole organization. Notice the paradox here. Being authentically purpose driven can be a powerful business strategy.

But you can't decide to be authentic because it will be good business. That wouldn't be authentic. Becoming authentically purpose driven is all about exploring the boundary between purpose and profit—about choosing to do the right thing and then fighting hard to find the business case to make it possible.

Mark made important strides in remaking Aetna, but its fate as a purpose-driven business is now in the hands of CVS, which bought the company in 2018, partly in the hope that Aetna's new strategy would complement CVS's desire to make its retail pharmacies centers for neighborhood health care and partly—perhaps—because CVS is itself experimenting with purpose.[21] But his experience illustrates both the power of purpose to seed the kind of architectural innovation that could mark the beginning of new forms of shared value creation across US health care, and how building a genuinely purpose-driven organization can itself be a strategy to create shared value.

Effective purpose-driven organizations share two elements. The first is a clear sense of their mission in the world. While the leaders of purpose-driven firms are very much aware that they must generate profits to survive, making money is not their primary goal. Some purpose-driven firms exist to improve the lives of their customers. Some focus on creating employment. Others hope to solve the world's environmental and social problems. But in every case they put mission over the need to maximize short-term shareholder returns.

The second element is a commitment to building an organization in which every employee is treated with dignity and respect and viewed as a whole human being whose autonomy and worth is to be honored. In these "high road" or "high commitment" organizations, authority is broadly delegated and work is designed to empower people on the front line to make decisions and improve performance. People are routinely challenged and given opportunities for personal growth. High commitment organizations pay well, but rely more on intrinsic motivation than on the use of monetary rewards or the threat of termination. Hierarchy is downplayed in favor of the development of trust and mutual respect between superiors and employees.

It is the combination of mission and this change in the nature of work that releases the creativity, commitment, and raw energy that enables purpose-driven firms to survive in a ruthlessly competitive world—and that drives the innovation that's required to reimagine capitalism. Leaders who embrace a mission without changing the way they manage often find themselves struggling to implement it. Those who simply raise wages without changing the nature of work and the purpose of the organization find themselves struggling to afford the pay raise. In short, the creation of authentically purpose-driven, high road organizations is an important step toward a just society.

While there is healthy job growth at the high end of the income distribution, the jobs that have traditionally provided a route to the middle class—in manufacturing and in entry-level clerical and technical work—are disappearing. The new jobs are either essentially temporary or are in fields like health care and elder care. Generally—except in the presence of strong unions—these are terrible jobs: paying badly, offering no benefits, and often mandating erratic and arbitrary schedules. Having a good job is fundamental to most people's sense of well-being. It is almost impossible to have a good life if your basic needs for food, shelter, and security are not met, but good jobs are also sources of social status, companionship, and a sense of meaning that greatly increase happiness.[22]

Does it sound as though I've drunk too much purpose-flavored Kool-Aid, and this is all merely happy talk? The rest of this chapter tries to persuade you that on the contrary, I am deadly serious. There is a great deal of evidence that purpose-driven firms not only routinely survive under brutally competitive conditions but also that they often significantly outcompete their more conventional rivals.[23] I begin by providing a sense of what effective purpose looks like in practice to illustrate why and how it can be a viable strategy. I then answer the important question of why—if this is such a great way of managing—it hasn't already taken the world by storm. I close by exploring why it is the case that more and more firms are announcing their commitment to purpose.

Purpose in Practice

The way in which the combination of mission and the nature of work plays out in practice to support both architectural innovation and the generation of great jobs can be seen particularly clearly at King Arthur Flour ("KAF"), the oldest flour company in the United States.[24] KAF's best-selling product, the five-pound bag of unbleached, all-purpose flour, is not a sexy product, and the market has been shrinking for years. Fewer and fewer people bake, and more and more flour is bought online, where brands often carry little weight. But KAF is thriving. Its customers love the company. KAF has over a million likes on Facebook and more than 375,000 followers on Instagram.[25] (For comparison, General Mills, the current market leader with $3.9 billion in sales in "meals and baking," compared to KAF's roughly $140 million, has about 85,000 likes on Facebook and 3,000 Instagram followers.[26]) Sales are growing in the high single digits annually—an unheard of growth rate for a commodity product in a two-hundred-year-old industry.

KAF's purpose is to "to build community through baking,"[27] and the three co-CEOs (!) have a very clear sense of just why and how home baking can make a difference in the world. Karen Colberg, the chief brand officer and one of the co-CEOs told me:

> Baking uniquely enables people to unplug. And as a mother of three teenagers, I'm constantly in this mode of wanting connection with my family and spending time together. And what we offer people is the ability to come together and do something.

Ralph Carlton, co-CEO and chief financial officer, put it this way:

> When you think of baking as opposed to food, you give people gifts. The emotional connection people have, the notion of the smell of fresh baked bread, there's something unique about baking that brings people together. And that inspires us . . . everything we do is centered around the baking experience.

Suzanne McDowell, co-CEO and vice president of human resources, added:

> Well, everyone can bake. So if you just start there, and you think about how baking can level the playing field—it doesn't matter how smart you are, or how wealthy you are, or any of the things that separate us: we can all come together and bake together. And spending time with people, baking and learning a life skill, no matter whether you are young or old, can be a remarkably unifying experience. You can bake with family, coworkers, and neighbors. Baking is an amazing opportunity to build community. And we need community building. It's really important in our world, always has been, always will be.

As in the case of Aetna, the passionate embrace of this purpose has enabled KAF to identify a strategy that is classically architectural. KAF no longer thinks of itself as selling only white flour—instead it is selling an experience, and supporting its customers in becoming great bakers. In Ralph's words:

> One of the challenges of baking, but one of the great things about baking, is to bake well you do need knowledge. And often you need inspiration. Very few people bake without reaching for a recipe or some other guidance. And baking is not that forgiving. It isn't like cooking, where you just go in there, and it doesn't matter what you

> do, something relatively good comes out. In baking, you have failure. [So] we started providing information on the web. And it has really grown from it being a small part of what we do to us being one of the leading sources of knowledge and inspiration for bakers around the country right now.
>
> And that's a core piece of our strategy . . . we're making a big bet that future generations of bakers, when they have to choose products, are going to choose the products from companies where they learn the most from and they trust the most. And it's not going to be because I barked at you and told you to go buy King Arthur. It's going to be because we had a great recipe, or we taught you a technique that you value very highly . . . [because] King Arthur is really a company that cares about me and cares about baking and cares about quality.

This strategy is enabled by a deeply participative, fully empowered workforce that embraces it as a reason for working that extends far beyond a paycheck—and that makes it immensely difficult to imitate. KAF's Vermont headquarters—now a major tourist attraction[28]—includes a retail store, where visitors can watch baking demonstrations and sample baked goods (made with KAF products, of course) and a baking school, where hundreds of passionate bakers arrive to take classes from King Arthur's master bakers. The company also offers online recipes and baking classes, and a fully staffed baking hotline, where customers can get answers to their baking questions from employees with thousands of hours of baking experience.[29] Everyone is passionate about baking. Everyone goes the extra mile to help the company succeed. The latest financial results are shared with every employee, and everyone is offered training on how to read income statements and balance sheets. The company is very careful about the people it hires, and then equally careful about how they are treated. Karen expands:

> The culture is a very present part of the hiring process. So when we meet people, and we talk about coming to work for King Arthur Flour, we talk about it being participatory. We talk about it being collaborative. But what does that mean? I want people to show up and feel accountable for themselves, accountable to their teams, and to have a clear understanding of what they're supposed to be doing. Also to be comfortable that they can challenge what it is they're doing and challenge what others are doing. And ask us questions so we have a really productive dialogue around issues such as: Where's the company going? Why did you decide to do that? Did you think about this?

Ralph adds the following:

> It's a culture where people reach inside themselves to do the right thing. Karen often gives the example, during our holiday season when business is crazy and we're sending thousands and thousands of packages out of our distribution center every day. Word spreads around the building that there's too much work down in Pick and Pack and the team needs help. And people just do it, they come downstairs to lend a hand, and not because the boss tells them to.

Suzanne also commented on the positive work environment:

> People are engaged. They're proud of our products. They are in it together. It's not like you're siloed, and here I am in my space. I'm going to do my job—it has no impact or effect on you. In fact (your job) has a lot of impact and effect on everyone. It's fun. We love to celebrate. We love to bake. We generally are pretty psyched about coming to work every day.

KAF's competitive success is thus intimately linked to its willingness to empower its workforce—and this empowerment in turn means not only that it's fun to work at KAF but that the company can pay over the odds and offer employees who vest a chance to build retirement savings. (KAF is a completely employee-owned company, which has potentially important implications that I will come back to in the next chapter.)

Creating a strong shared sense of purpose in a relatively small firm like King Arthur Flour is one thing. Can it be implemented in much larger organizations? It can. Toyota's example underlines the fact that it's possible to create a similar sense of creativity and commitment in a billion-dollar organization that employs hundreds of thousands of people.

Toyota is a deeply purpose-driven organization. The Second World War destroyed the Japanese economy and most of the infrastructure and the housing stock. In 1950 Japan had a GDP less than half that of Norway or Finland, despite having a population roughly twenty times the size of either.[30] In this context, Toyota's leaders—like those of many of the most successful Japanese firms of the time—had two goals: to create employment and—since Japan has almost no natural resources—to build thriving enterprises that could compete at an international scale. The firm was founded in 1937, but in 1950 a crippling labor dispute nearly forced it into bankruptcy. Desperately in need of cash and teetering on the edge of failure, Toyota was able to use this profound commitment to the community to translate the threat of bankruptcy into a new way of working that was an order of magnitude more productive than the management style of its American competitors, as well as an enduring source of great jobs.[31]

When in 1957 Toyota first opened an office in the United States, General Motors produced one in every two cars sold in America. GM's executives laughed at the Japanese imports, confident that they had a lock on the American consumer. But by the 1980s American consumers had fallen in love with Japanese cars. They complained that American cars suffered from noise and vibration, and that they were significantly less reliable than their Japanese competitors. Toyota was able to build much better cars for about the same price by thinking quite differently about the system through which cars were designed and produced, and completely changing the relationship between the key actors in the system—in short, by completely "re-architecting" how cars were designed and built.

Work at GM—as at nearly every other company in the United States at the time—had historically been organized along strictly functional and hierarchical lines. Responsibility for the design and improvement of the assembly system was vested firmly in the hands of supervisors and manufacturing engineers, while vehicle quality was the responsibility of the quality department, which inspected vehicles as they came off the assembly line. GM's managers were notorious for believing that blue-collar workers had little—if anything—to contribute to the improvement of the production process. Workers would perform the same set of tasks—for example, screwing in several bolts—every sixty seconds for eight to ten hours per day. They were not expected or encouraged to do anything beyond this single task. Relationships between those working on the plant floor and local management were actively hostile. One worker interviewed in the early nineties described life then this way:

> In the old days, we fought for job security in various ways: "Slow down, don't work so fast." "Don't show that guy next door how to do your job—management will get one of you to do both of your jobs." "Every now and then, throw a monkey wrench into the whole thing so the equipment breaks down—the repair people will have to come in and we'll be able to sit around and drink coffee. They may even have to hire another guy and that'll put me further up on the seniority list."
>
> Management would respond in kind: "Kick ass and take names. The dumb bastards don't know what they're doing." . . . Management

> was looking for employees who they could bully into doing the job the way they wanted it done. The message was simply: "If you don't do it my way I'll fire you and put somebody in who will. There are ten more guys at the door looking for your job."

GM had an analogous relationship with its suppliers, treating them as interchangeable and driving down costs by pitting them against each other. But Toyota demonstrated that it was possible to structure work in significantly different ways. Jobs on Toyota's production line were even more precisely defined than those on GM's: for example, detailed instructions for every station specified which hand should be used to pick up each bolt. But Toyota's employees had a much broader range of responsibilities.[32] Each worker was extensively cross-trained, and was expected to be able to handle six to eight different jobs on the line. They were also responsible for both the quality of the vehicle and for the continual improvement of the production process itself. Everyone on the line was expected to identify quality problems as they occurred, to pull the Andon Cord that was located at each assembly station to summon help to solve problems in real time, and if necessary to pull the "Andon Cord" again to stop the entire production line. Workers played an active role in teams that were responsible for identifying improvements to the process that might increase the speed or efficiency of the line. As part of this process, workers were trained in statistical process control and in experimental design.

Supervisors and industrial engineers still existed at Toyota, but they were explicitly charged with being of service to the shop floor workers. Everything was in the service of the continual improvement of the process, and it was the workers on the floor who were in charge of improvement. Toyota had a strongly egalitarian culture, and "respect for people" was one of its core values.

The firm brought the same combination of deep respect and widespread empowerment to its relationship with its suppliers. Suppliers were treated as "supplier partners" and trusted with proprietary information that enabled them to work closely with Toyota in the service of building better cars. Toyota even changed the nature of white-collar work in the industry. Inside the firm, people in marketing and engineering were encouraged to think of themselves as allies rather than as adversaries. The finance function was encouraged to support the process of continual improvement, rather than to act as a ruthless police force in the service of the bottom line. Employees were encouraged to view themselves as in service to the purpose of the firm, rather than in service to their own advantage.

This strategy was spectacularly successful. In the late 1980s the Japanese took 1.7 million engineering hours to develop a $14,000 car, while their US competitors took almost twice the time. On the production line, GM took nearly twice the number of hours, compared with Toyota, to assemble a car.[33] By 1990 Toyota's market value was twice that of GM's, and by 2008, the firm was the world's largest automobile producer.[34]

Let me put this another way: Toyota was developing new cars in half the time and at half the cost of its American rivals and building them using half the number of people. It took American firms nearly twenty years to come to terms with these results, despite the fact that Toyota's success was widely documented. There have been at least three hundred books and more than three thousand academic articles written about the firm.

Moreover, Toyota is not unique. In every industry, on average the most productive firms are more than twice as productive as the least.[35] Recent research drawing on data collected from thousands of firms across the world has confirmed that these differences are almost certainly driven by the ways in which firms are managed.

"High commitment" work practices drive increases in productivity across an extraordinary variety of industries.[36]

If managing with purpose is not only possible but potentially such a powerful source of competitive advantage, then why doesn't everyone manage this way? Why have so many firms been so slow to put these ideas into action? Gallop reports that 34 percent of US workers are now "actively engaged"—the highest number in Gallop's history—and that the percentage who are "actively disengaged" has fallen to 13 percent, a new low. But more than half of all employees remain "unengaged"—generally satisfied but not cognitively or emotionally connected to their work or workplace. They show up and do the minimum that is required but are likely to leave if they receive a slightly better financial offer.[37]

The reason purpose-driven management is not universal or at least more common is because it is *in itself* an architectural innovation of the first order—requiring managers to think about themselves, their employees, and the structure of the firm in entirely new ways. And unfortunately, many managers are prisoners of a worldview—and with it a view of employees and a method of management—that is over a hundred years old. If we are to reimagine capitalism, it is vital to understand just where this worldview came from—and how it can be changed.

Shaping Worldviews for a Hundred Years

The great Victorian-era capitalists viewed their employees as fundamentally selfish and lazy, motivated mostly by money, and needing to be carefully controlled. Firms were run along strictly hierarchical lines, with a strict division between management and employees and an almost uniform assumption that labor and capital were destined to be in conflict. Businessmen of the time generally assumed that building a successful business required tightly supervising the workforce and keeping wages as low as possible. In the United States they broke unions when they could, hired private armies to fight—and kill—striking employees, and persuaded the US Supreme Court that unions should be prosecuted under the antitrust laws.

This view of the vast majority of employees as essentially stupid machines that were best harnessed by the skill and expertise of managers was strongly reinforced by the invention of "scientific management," a perspective that gave a scientific imprimatur to the belief and that made it the conventional wisdom not just at GM but at most large firms for most of the twentieth century.

Scientific management was invented by a man named Frederick Taylor. (Indeed, the technique is often called "Taylorism.") Taylor was an American blue blood. He was descended from one of the Mayflower pilgrims, attended Phillips Exeter Academy, and was admitted to Harvard. But—perhaps because of rapidly deteriorating eyesight—he instead decided to serve a four-year apprenticeship as a shop floor machinist, ultimately joining the Midvale Steel Works in 1878 as a machine shop laborer. He was quickly promoted and ultimately became chief engineer of the works.

Taylor's experience in these roles convinced him that the vast majority of the plant's employees were "soldiering," or deliberately working as slowly as they could, and he began a systematic study of what we now call "productivity." He discovered that in many cases output could be greatly increased by breaking every action down into its constituent parts, improving the productivity of each part, and then forcing employees to follow precisely the procedures laid out for them by management. In practice this meant using the promise of financial reward to turn people into robots. One of his most famous stories is about the loading of pig iron at a firm called Bethlehem Steel. The experiment began with a man he called Schmidt. Taylor tells the story this way:[38]

> The task before us, then, narrowed itself down to getting Schmidt to handle 47 tons of pig iron per day and making him glad to do it. This was done as follows. Schmidt was called out from among the gang of pig-iron handlers and talked to somewhat in this way:
>
> "Schmidt, are you a high-priced man? . . . What I want to find out is whether you are a high-priced man or one of these cheap fellows here. What I want to find out is whether you want to earn $1.85 a day or whether you are satisfied with $1.15, just the same as all those cheap fellows are getting.
>
> Did I vant $1.85 a day? Vas dot a high-priced man? Vell, yes, I vas a high-priced man.
>
> Well, if you are a high-priced man, you will do exactly as this man tells you to-morrow, from morning till night. When he tells you to pick up a pig and walk, you pick it up and you walk, and when he tells you to sit down and rest, you sit down. You do that right straight through the day. And what's more, no back talk. Now a high-priced man does just what he's told to do, and no back talk. Do you understand that? When this man tells you to walk, you walk; when he tells you to sit down, you sit down, and you don't talk back at him. Now you come on to work here to-morrow morning and I'll know before night whether you are really a high-priced man or not."

Taylor goes on to explain how Schmidt was turned into a human robot—working when he was told to work, resting when he was told to rest—and how this discipline increased his productivity by more than 60 percent. This claim is almost certainly exaggerated.[39] But there's lots of evidence that the use of Taylor's methods dramatically increased productivity in a wide range of settings. Advocates of Taylor's approach still claim that putting all the expertise in the hands of management and managing people as if they were machines may have its downsides, but it has such dramatic effects on productivity that these costs are well worth paying. "Taylorism" became the conventional wisdom, and Taylor's *The Principles of Scientific Management* became the best-selling business book of the first half of the twentieth century. Taylor's ideas became so widely accepted that the early evidence that embracing a purpose could significantly improve performance was widely dismissed—and even when it was finally acknowledged, firms found it enormously difficult to implement the new ways of working. Take, for example, GM's struggle to respond to Toyota's success.

The Struggle to Find a New Way: GM Responds to Toyota

By the early 1980s GM's leaders had become convinced that Toyota was indeed doing "something different" in its factories. But they initially refused to believe that the essence of Toyota's advantage lay in its relationship with its employees. Instead they focused on tangible changes to the production process—tools like the fixtures designed to change stamping dies rapidly, or the use of "just in time" inventory systems—rather than on the set of management practices that made it possible to develop and deploy these techniques. For example, one consultant to GM in the 1980s reported:

> One of the GM managers was ordered, from a very senior level—[it] came from a vice president—to make a GM plant look like NUMMI [a plant that Toyota had taken over from GM and completely transformed]. And he said, "I want you to go there with cameras and take a picture of every square inch. And whatever you take a picture of; I want it to look like that in our plant. There should be no excuse for why we're different than NUMMI, why our quality is lower, why our productivity isn't as high, because you're going to copy everything you see. . . ." Immediately, this guy knew that was crazy. We can't copy employee motivation; we can't copy good relationships

> between the union and management. That's not something you can copy, and you can't even take a photograph of it.[40]

Performance at General Motors was judged on the basis of well-defined rules or easily observable metrics, such as whether individuals met prespecified deadlines, while performance at Toyota was judged on the basis of the performance of the team as a whole.[41] At Toyota, goals were jointly determined through lively communication across multiple levels of the organization, an idea completely foreign to the top-down, command-and-control manner in which GM was run.

Managers whose entire careers had been spent focusing on the current quarter, making their numbers, and learning to fine-tune an existing system were ill-equipped to rethink the fundamentals of employee management. People who had always managed their suppliers and their blue-collar workers by bullying them had a difficult time thinking of them as a source of continuous improvement and treating them with trust and respect.

Most critically, the successful adoption of high-performance work practices requires the ability to create deep levels of trust, and GM's history meant that it had horrible problems doing this. GM preferred to manage by the numbers and to promote on the basis of quantitative results. But no set of numerical objectives can specify the kinds of behaviors that characterize high-performance firms. Senior management would announce a commitment to long-term relationships and to building trust, but until and unless these announcements were coupled with similar commitments and altered incentives at the local level, few believed that the local managers whom they dealt with would, in fact, change their behavior.

An extraordinary example of the value that can be created by this kind of trust is that for many years, Nordstrom's employee handbook was a single sheet of paper on which was written:[42]

WELCOME TO NORDSTROM

We're glad to have you with our Company. Our number one goal is to provide outstanding customer service. Set both your personal and professional goals high. We have great confidence in your ability to achieve them.

Nordstrom Rules: Rule #1:
Use good judgment in all situations.

There will be no additional rules.

Please feel free to ask your department manager, store manager, or division general manager any question at any time.

They meant it. Nordstrom built an impressive track record in the retailing business on the basis of their employees "using good judgment." In a series of famous cases, one sales associate accepted the return of snow tires (Nordstrom does not sell snow tires), another drove for hours to deliver a set of clothes so that a customer could attend a family occasion, and a third changed the tires of customers stranded in the company parking lot. Stories like these gave Nordstrom a reputation for excellent customer service that was the envy of its competitors and that created deep customer loyalty.[43]

But managers will only trust employees to "use their good judgment" if there's a long history of it working out well—and employees will only take the risk of acting on their own initiative if the firm has a long history of rewarding them if they do. Real trust can only be built over time—and only by firms that are willing to make the short-term sacrifices that are required in any relationship to demonstrate authentic commitment.

GM's obsession with short-term returns and numerical targets made it enormously difficult to build this kind of trust. In 1984, for example, the company announced that it was interested in modifying the union contract to support the use of teams and joint problem solving—but then a leaked internal memo suggested that GM was planning to use the new contract simply to reduce head count. Throughout the 1980s, many in union leadership remained convinced that GM was implementing Toyota's practices only as an attempt to speed up production and to put employees under even greater pressure. GM thus faced significant problems in building trust within its workforce. It's hard to trust an entity whose avowed purpose is to make money at your expense at every possible moment.

It's tempting to believe that GM was just uniquely badly managed, but I have seen similar problems crop up in firm after firm attempting to adopt high-performance work practices. Many managers are reluctant to give up the comforting assumption that employees are stupid and managers hold all the cards. They don't want to do the hard emotional and mental work of building a way of working in which everyone is honored and power is widely distributed. And when they do decide to take the step, it can be hard to make the long-term investments that are critical to building trust.

You can see this dynamic playing out over 150 years of history. Purpose-driven firms emerge—demonstrate the power of purpose-driven management—and are dismissed. But it is their experience that has laid the foundations for our current moment, building a deep body of experience that remains relevant today.

The Rise of "High Road" Firms

In 1861 when George and Richard Cadbury took over their father's failing tea and coffee business, they built one of the first explicitly purpose-driven firms into one of the most successful firms in England.[44] Their deep commitment to a faith tradition that stressed the equality of all human beings was instrumental to their success, highlighting the critical role that strong spiritual or political convictions often play in giving leaders the courage and vision necessary to manage with purpose.

The Cadbury brothers were born in Birmingham, in the United Kingdom, from three generations of Quakers, or as they preferred to be called, members of the Society of Friends. Members of the Society were deeply committed to "the light within," or to the belief that God manifests directly in everyone, and were prominent in the fight against slavery and in campaigns for penal reform and universal education. As a community they were suspicious of profit, believing that the function of industry should be to serve the community as a whole, and that conflict between labor and management should be resolved through open conversation and goodwill.

The two Cadbury brothers inherited a firm that was in trouble. It had only eleven employees (down from twenty) and was losing money. Together they invested the £8,000 they had inherited from their mother (roughly £700,000/$861,300 in today's money)[45] and began the hard work of turning the business around. By 1864 they were showing a small profit, and over the following decades they built one of the most successful firms in England. Their purpose was straightforward: to sell cocoa and chocolate of the highest possible quality. When they began, the products on the market were basically a highly diluted gruel. According to George Cadbury "only one fifth of it was cocoa, the rest being potato starch, sago, flour and treacle." Cadbury's Cocoa Essence hit the market in 1866, supported by the slogan "Absolutely pure, therefore best" and quoting physicians in its advertisements. In 1905 the firm launched "Cadbury's Dairy Milk," again stressing its purity since it used fresh rather than powdered milk. A commitment to using the best possible ingredients would remain a hallmark of the firm for the next century. Like all high-performing purpose-driven firms, they combined a strong sense of the social purpose of the business with a radically different way of treating their employees.

In 1878 the brothers built a large factory about four miles outside Birmingham. The factory—which was given the name "Bournville" in an attempt to suggest a French association for the chocolate—thrived, and in 1895 the Cadburys bought an additional 120 acres and set out to build a model village around it full of trees, flowers, and gardens. George was an active teacher in the Quaker Adult School Movement and had spent years teaching in Birmingham's worst slums. The experience led him to realize that it was as important to transform the living conditions of the poor, as it was to provide them with an education. "If I had not been brought into contact with the people in my adult class in Birmingham," he later said, "and found from visiting the poor how difficult it was to lead a good life in a back street, I should probably never have built Bournville village." This stress on personal experience with those who are less privileged than oneself is another theme that recurs in the accounts of purpose-driven leaders: hands-on experience often provides a motivation for exploring a different way of leading.

From the beginning the brothers explicitly rejected Taylor's approach to management. "Even if on the productive side," one of the brothers remarked in a 1914 paper entitled "The Case Against Scientific Management," "the results are all that the promoters of scientific management claim, there is still the question of the human costs of the economies produced." George Cadbury told the Conference of Quaker Employers that "the status of a man must be such that his self-respect is fully maintained, and his relationship with his employer and his fellow-workmen is that of a gentleman and a citizen."

The brothers treated their employees as family and were famous for their reluctance to stand on ceremony and their willingness to get their hands dirty. From the beginning they invested heavily in education. All employees were required to take an introductory academic course and could then choose commercial or technical training if they wished. The Cadburys provided sports and physical education facilities. They introduced sick pay and a pension fund. There were Christmas, New Year's, and summer parties. They also experimented with worker participation in the running of the plant. The firm was formally run by the Board of Directors (all of whom were family members) but controlled on a day-to-day basis by a series of committees, including a Men's Works Committee. The Works Committee included both staff and foremen and was responsible for factory conditions, quality control, and welfare work. In 1902 Cadbury introduced a suggestion scheme through which directly elected employee representatives were invited to suggest improvements in the management of the plant, and in 1919 the company began to experiment with full-fledged industrial democracy, creating a three-tiered structure of Shop Committees and Group Committees reporting to a Works Council.

By the 1930s Cadbury was the twenty-fourth-largest manufacturing company in England and had created a portfolio of brands that remain global powerhouses today. (Cadbury's Fruit and Nut Bar was a staple of my childhood, and even now, one of my life's pleasures is quietly consuming one every time I'm in England.) It remains an inspiration to many purpose-driven leaders, but at the time Cadbury's experience was put down as an oddity—a function of the owners' Quaker beliefs, rather than of a new method of management.

The idea that there might be a better way to manage continued to surface. In the 1940s Eric Trist, a British academic, was invited to visit Haighmoor, a British coal mine about fifty miles east of Cambridge. The vast majority of British coal mines were conventionally organized along Taylorist lines, but Haighmoor was different. At Haighmoor conventional equipment couldn't access the coal face, and the miners had instead developed a system that used self-organized teams in which every miner might handle up to six different jobs.

The mine was far safer and far more productive than any of its competitors', but when Trist suggested that it might be a good idea to use some of Haighmoor's techniques in other mines, the government agency that had let Trist study Haighmoor in the first place rejected the idea. The agency apparently feared that Trist's interference would only lead to unrest and forbade him to include the name "Haighmoor" in his reports. The fear that empowering employees would lead to uppity employees and significant friction between labor and management continued to surface as a reason to reject high road management, despite the fact that experience suggests precisely the opposite is true.

Trist next began collaborating with a group of Norwegians who had fought Hitler successfully in small, self-managed teams and had together set up teams across Norwegian industry. He believed their approach had the potential to drive dramatic improvements in performance. However, while he was able to interest a few corporations in his ideas, sooner or later, most managers retreated queasily, "as if they had just discovered that Trist was trying to kill them off." In a sense, of course, he was. "Their opinion had a modicum of correctness in it," he said, years later. "They'd had all the power and did what they liked and they didn't want to share power."[46] As we saw in the GM case, this remains an important barrier to change. Successful purpose-driven leaders have to learn to relinquish control—to believe at a deep level that those who work for them have at least as much creativity and drive as they do.

Trist's ideas resurfaced in the work of Douglas McGregor, a professor at MIT's Sloan School of Management. In *The Human Side of Enterprise*, published in 1960, McGregor laid out two theories of human motivation, foreshadowing much of modern motivational theory. The first, "Theory X," framed people as fundamentally selfish and lazy, willing only to work for themselves and for extrinsic rewards such as money, status, and power. The second, "Theory Y," hypothesized that people are motivated as much or more by intrinsic rewards—by the pleasures of mastery and autonomy, by the opportunity to build relationships with others, and by the desire for meaning and purpose. Theory Y anticipated much modern research in postulating that people are as much "groupish" as they are selfish, that they are hard wired to enjoy being part of a group and even—under certain circumstances—to act cooperatively and even altruistically. The book was sometimes interpreted as an argument in favor of Theory Y, but McGregor himself insisted that his point was not that Theory Y was correct, but that both theories are useful models, and that to rely on Theory X alone is a dangerous oversimplification that leaves many powerful sources of motivation off the table.

One of the first groups to put McGregor's ideas into action were managers in Augusta, Georgia, who were making laundry detergent for Procter & Gamble (P&G).[47] P&G had been an early and enthusiastic adopter of the principles of scientific management, but by the early 1960s, the managers building the Augusta plant were becoming increasingly frustrated by its limitations. Everything was rigorously measured. Everything was specified. Everything was wrapped in a tangle of rules and procedures. They decided to try something different.

They began by inviting McGregor to visit. They liked his "plain speaking, brutally frank, fiercely engaged management style" and his description of Trist's techniques, and decided to put them into practice. They moved to a system under which the Augusta plant was run entirely by "technicians" organized into teams, each of whom was expected to develop a wide range of skills and to actively contribute to the continual improvement of the plant. The plant had no job classifications and no production quotas. Its employees spent four hours a week in training and an additional two hours meeting together to solve problems. In short, the plant invented something remarkably close to the Toyota production system years before Toyota would first make waves in the United States. Augusta was so successful that by 1967 every new P&G plant was required to use the system.

The first plant designed from the ground up to use the new techniques was built in Lima, Ohio. Under the leadership of Charlie Krone—an unconventional plant manager who had studied not only Trist but also Tibetan and Sufi mysticism and the work of the spiritual teacher George Gurdjieff—the Lima plant was designed to "embody learning" and to integrate emotional and psychological factors directly into the design of the work. The human needs of the employees were considered to be at least as important as those of the business. There was minimal hierarchy. If you wanted to tackle an issue, you worked to persuade your colleagues to build a team around you. Teams managed their own schedules while managers acted as coaches and enablers. It too was enormously successful. Production costs were rumored to be half the costs of a conventional plant, but may have been even less than that. The plant's managers apparently thought that nobody would believe the real figures.

But the managers from Lima who attempted to persuade others to use the techniques they had pioneered met with very little success. Senior managers were at first puzzled and then threatened by the "hippie talk" of those who had experienced the new way of doing things. In turn the new leaders became deeply frustrated with their superiors, further persuading senior leaders that they weren't really serious and that they couldn't be trusted. Krone became an outsider, a management guru working with a tight cadre of apprentices to spread the new techniques in smaller firms outside the Fortune 500.

The idea that empowering employees, treating them with trust and respect, and motivating the organization through common goals and shared purpose could dramatically increase performance did not die. Trist became one of the founders of the Tavistock Institute, a group of researchers that stressed the importance of human relations in shaping work. Douglas McGregor's work at MIT influenced the work of his colleague Ed Schein, who became the world's foremost expert on organizational culture. Scholars like Michael Beer at Harvard continued to write about high-performing firms whose success was rooted in purpose-driven leadership and respect for employees. But for decades the purpose-driven organization was an outlier, more the exception than the rule.

However the world has changed dramatically in the last ten years. The idea that purpose can drive performance has become something close to the conventional wisdom. In one survey, four out of five CEOs agreed that "a company's future growth and success will hinge on a values-driven mission that balances profit and purpose" and that "empowering employees' personal sense of purpose and giving them opportunities for more purpose-driven work is a win-win that is good for both their business and the employees themselves."[48]

There are many reasons for this shift. Reputational management and sheer expediency are certainly playing a role. More and more companies feel the need to show that they are doing "something," and many have come to recognize that a shared purpose is a great tool for driving transformation and growth.

But there is also an enhanced appreciation of the roots of Toyota's success. Like GM, many firms initially interpreted Toyota's exceptional performance as a reflection of the firm's adoption of high-performance work practices—for instance, a reliance on teamwork and a focus on incremental innovation—rather than of the organizational culture and values that made this adoption possible. As firms have struggled to adopt these practices themselves, this has begun to change. In the words of one of the organizational consultants whom Toyota employs to teach other firms about the Toyota way, "It's all about the culture. It was always all about the culture. But it takes firms a long time to fully take this on board."[49] The outstanding success of other employee-orientated companies like Southwest Airlines and Whole Foods has also attracted widespread attention.

Another force behind the recent shift is the outpouring of new research linking the adoption of authentic purpose and high-performing work systems to financial performance. In her book, *The Good Jobs Strategy*, for example, MIT researcher Zeynep Ton show that leading retailers like CostCo and Mercadona, which redesigned their operations to support continuous learning and employee initiative, were able outperform their competitors and to pay their employees significantly over the odds. Other academic researchers using many years of data and fine-grained measures of purpose have showed that high levels of employee satisfaction and the presence of a purpose closely linked to strategy improve total shareholder returns.[50]

The world is also changing in ways that make the need for purpose increasingly evident. Public expectations are shifting, with 73 percent of the world's population now expecting business to address the big problems of our time.[51] The millennials and their successors are actively seeking jobs that support a sense of meaning and purpose. At the same time, the trust gap between business and the general public is accelerating. A third of employees don't trust their employer, and while 82 percent of the elite trusts business, only 72 percent of the general public do.[52]

Something more profound is also going on. As the problems we face become increasingly pressing, many business leaders are recognizing that there is a clear moral imperative for action. I have talked to hundreds of business leaders about their strategies for creating shared value. They have all been eloquent about the business case. But in private, in the corridor or over a beer, nearly all of them have told me that it is a compelling purpose that drives them to act—that it is the existential risk that climate change presents, or the need to rebuild their community, or to transform health care, or to save the oceans that drives them forward. Some people interpret this duality as hypocrisy. But I believe that it is fundamental to reimagining capitalism, and that business leaders must be aware of the need to make both profits and meaning if we are to solve the great problems of our time. Purpose-driven leadership is essential if we are to uncover the new business models that will create shared value, if we are to be able to implement them, and if we are to create the good jobs and the kinds of workplaces that are essential to building a strong society.

Yet many firms struggle to integrate purpose into their operations. In many firms purpose has yet to be clearly specified, linked to strategy, or communicated to employees.[53] Part of this disconnect undoubtedly reflects a deep unwillingness to let go of Taylorism as a philosophy or to appear emotional or sentimental at work. But there is another, more structural barrier at work—and that's the short-termism and ignorance of the world's investors. When investors insist on steadily increasing quarterly earnings and don't understand or cannot measure the value of purpose, it can be very hard to make the long-term investments required to become a purpose-driven firm.

That's why the third step toward reimagining capitalism is rewiring the capital markets.

5

REWIRING FINANCE

Learning to Love the Long Term

> If money go before, all ways do lie open.
>
> —WILLIAM SHAKESPEARE, *THE MERRY WIVES OF WINDSOR*

If the opportunities for building thriving companies that are not only highly profitable but also making significant progress against the big problems are real, and if becoming a values-based, purpose-driven firm is the royal road to unlocking these opportunities, why aren't more firms actively embracing the combination of purpose and shared value? Even as 85 percent of the world's largest companies claim to have a purpose, and many are beginning to explore what they can do to create social value, we are still a long way from this approach being anything like business as usual. What is going on?

The business leaders I know have an easy answer to this question: They say that even when their company wants to improve its social and environmental performance, they are constrained by our obsession with the short term. "It's the investors," they tell me, "they're obsessed with short-term results. It's impossible to invest for the long term without getting slammed if it means missing quarterly earnings targets." Peter Drucker, perhaps the most famous management guru of the first half of the twentieth century, memorably suggested: "Everyone who has worked with American managements can testify that the need to satisfy the pension fund manager's quest for higher earnings next quarter, together with the panicky fear of the raider, constantly pushes top managements toward decisions they know to be costly, if not suicidal, mistakes."[1]

Every CEO I've ever met agrees with Drucker. We know that companies routinely delay or eliminate profitable investment opportunities to ensure they hit their numbers. In one survey nearly 80 percent of chief financial officers said they would decrease spending on research and development to meet earnings targets, and just over 55 percent said they would delay a new project to meet earnings targets even if it meant a (small) sacrifice in value. In another, 59 percent of executives would delay a high net present value project if it entailed missing earnings by a dime.[2]

There's also reason to believe that while asset *owners* may be focused on the long term, asset *managers* may not be. Most asset owners do not manage their own assets. For example, in 2016, institutional investors held 63 percent of the outstanding public corporate equity.[3] The retirement assets of most pensioners are managed by pension funds, which in turn rely on professional asset managers to invest on their behalf. Most individual investors invest in mutual or index funds, where their assets are managed by professional asset managers—who also vote their shares. This means that the interests of the asset owners are not necessarily reflected in the behavior of those who actually manage the assets: while many asset owners might like to drive performance over the long term, many investment managers might well prefer short-term returns, particularly if their compensation or the size of their portfolio is shaped by their ability to deliver immediate returns.[4]

When in October 2015, Doug McMillon, the CEO of Walmart, announced that Walmart's sales would be flat for the year and that earnings per share would fall 6–12 percent, the value of Walmart's stock sank by nearly 10 percent, taking with it roughly $20 billion in market value.[5] McMillon had attempted to explain that the decline in earnings reflected a $2 billion investment in e-commerce and a nearly $3 billion investment in paying hourly employees more—both moves that he believed were essential for the health of the business—but Wall Street was not impressed. Walmart's stock is still majority-owned by the Walton family, who were strongly supportive of the decision, so Doug kept his job, but many CEOs fear that in similar circumstances they would not be so fortunate.

Many business leaders tell me they are reluctant to jump wholeheartedly into embracing shared value because the need to satisfy their investors and to avoid the threat of activist interest in the stock makes it impossible to invest in the kind of long-term projects that true purpose requires. The way to reimagine capitalism, they suggest, is to get investors off their backs.[6] They propose a variety of ways in which this might be done—from changes in the law that would make it clear that firms have responsibilities to multiple stakeholders, to only allowing investors to vote their shares if they have held them for a sufficiently long period of time—but all of them are clear that if we are to reimagine capitalism, investors should have less power.

I'm very sympathetic to this argument. I, too, know of great projects that were delayed because they would harm next year's operating earnings. But I think that both the problem and the solution are more complicated than this line of reasoning would suggest.

There are at least two problems with the simple version of the short-termism thesis. The first is that while investors do punish firms that miss their earnings, the research in the area overwhelmingly suggests that this is because investors believe that missing earnings is an indication of bad management, rather than because they don't support investing in the long term.[7] Indeed one of the strongest findings in the accounting literature is that firms that miss their earnings do indeed perform worse—in the long term—than firms that make their targets.[8] So one interpretation of the drop in Walmart's stock price is simply that Doug's announcement that he was going to miss his targets led his investors to worry that something was fundamentally wrong with the firm—or with his leadership.

The second is that we know that in some circumstances investors are more than willing to invest in firms that won't be profitable for many years. Gilead, a biotechnology firm that is famous for introducing the first drug to cure hepatitis C, lost $343 million over the course of the first nine years after its IPO.[9] But it was valued at $350 million the year it went public, and nine years later was worth nearly $4 billion.[10] In the five years after Amazon first listed on the Nasdaq, it posted a cumulative net loss of just under $3 billion. But that year investors valued the company at just over $7 billion. Fourteen years later, when the company first moved solidly into the black, Amazon was worth $318 billion, despite the fact that it only had $600 million in profits.[11] Clearly many investors have been willing to wait for years to see Amazon's investments pay off. Indeed investors have been willing to funnel billions to a wide range of "platform" plays—including Uber, Lyft, and Airbnb—despite the fact that many of these firms have yet to make any money.

So it can't be that investors are altogether and overwhelmingly short term focused. When investors understand the nature of the bet they are being asked to make, some of them will make it. It took investors many years to learn the language of biotech. But now hundreds of analysts understand why it is that investing in fundamental research might yield billions of dollars in profit later. It took the dot-com bubble of 1994 to 2000, and the success of Facebook and Google to persuade mainstream investors of the power of platforms. But most now have a deep appreciation for the ways in which investing to build a large, sticky customer base committed to your platform can lead to massive profits. Indeed one of the causes of the recent failure of WeWork's IPO was the fact that most investors could see that while the firm touted itself as a platform play, it was not at all clear this was the case.

Viewed from this perspective, to the degree that there is genuinely money to be made in creating shared value and in building purpose-driven firms, investors' reluctance to invest in purpose cannot be only a function of their inherent short-termism. It must be—at least partially—a failure of information. Walmart's stock might have fallen on Doug's announcement because investors had no idea how to measure the impact of the investments he was making, and as a result simply didn't believe that they were likely to increase long-term returns. It turns out that reimagining capitalism requires reimagining accounting.

Give Investors Better Data

It took me a surprisingly long time to embrace the idea that accountants hold the key to saving civilization. Even after I'd read Jacob Soll's wonderful book, *The Reckoning: Financial Accountability and the Rise and Fall of Nations*—a blow by blow account of how the invention of double entry bookkeeping enabled the creation of the modern state—I secretly thought of accounting as the dusty, dry part of business—about as interesting as plumbing.

But then I noticed a strange thing. I knew plenty of business-people who were only mildly worried about the fact that we were running our entire economy on the basis of the idea that generating massive amounts of CO_2 typically doesn't cost firms a dime, or that it was equally costless (to the firm) to hollow out a community, pay one's employees bottom dollar, and push for tax cuts. But the accountants I knew weren't mildly worried. They were incandescently worried. We all have a general sense that what gets measured gets managed. But the accountants had spent their professional lives writing about how even tiny changes in accounting rules can change behavior in profound ways. And they could see that we were failing to measure a whole world of things that were shaping the performance of the business but that were effectively invisible.

Take "reputation," for example. We know that it can have profound economic effects. We know that it takes years to build and can be destroyed in an instant. Or "corporate culture"—ditto. But there's no mention—and certainly no measure—of either in the financial reports. If you make your decisions by analyzing financial statements, as most investors do, there's an enormous amount of information you just don't see. And if you don't see it, you're tempted to think it doesn't exist or doesn't matter.

Modern accounting provides the foundation for modern capital markets. Few people would trust their savings to strangers without some assurance that they would be able to tell whether they were putting their investors' interests first—and it's impossible to know whether this is the case without accurate numbers that reflect the health of the business. We tend to take the existence of things like the balance sheet for granted, but modern financial reports are the result of a hundred-year struggle over exactly which numbers firms should report—and who should be responsible for making sure these numbers are accurate. Until the disaster of the Great Depression led to strong public demands for financial transparency and the formation of the Securities and Exchange Commission (the SEC),[12] American firms didn't routinely report much in the way of financial information. Here, for example, is the full text of Procter & Gamble's annual report for 1919:[13]

Office of the Procter & Gamble Company
Cincinnati, Ohio, August 15, 1919

To the Stockholders of the Procter & Gamble Company:

The total volume of business done by this Company and constituent Companies for the fiscal year ended June 30, 1919, amounted to $193,392,044.02.

The net earnings for the year, after all reserves and charges for depreciation, losses, taxes (inclusive of Federal and State Income and War Taxes), advertising and special introductory work had been deducted, amounted to $7,325,531.85.

We shall take pleasure in furnishing further information to any accredited stockholder who is interested, and who will apply, in person, at the Company's office in Cincinnati.

Yours respectfully,
The Procter & Gamble Company
Wm. Cooper Procter, President

If you were a Procter & Gamble shareholder in 1919 and wanted to know more about your company than its annual revenues and profits, you had to go to Cincinnati and inquire in person. This made it very difficult to value companies unless one knew them very well indeed, which in turn limited the number of investors who were willing to invest in any single company. Modern financial accounting, in contrast, allows investors thousands of miles away to compare one company to another using standardized, audited measures whose link to performance is widely understood—and this in turn means that nearly everyone can invest everywhere, greatly increasing the odds that well-run companies will be able to raise capital.

One crucial step toward persuading investors to invest in business models whose success relies on the discovery of new customer needs, reduced risk, and high commitment organizations is the development of reliable, standardized metrics of aspects of the firm's strategy and operations that haven't historically been included in the financial accounts. Take "risk" for example. We know that climate change poses profound risks to some firms. But which firms? As consumers and governments wake up to environmental and social problems, firms that make their living pumping greenhouse gases into the atmosphere or selling products made in abusive labor conditions are at risk. But you can't look at their financial accounts and know how much they're pumping out or whether their suppliers are violating human rights.

Or take "culture." Many people have a sense that corporate culture can be a profound source of long-term advantage and that choosing to treat employees well could greatly increase the productivity of the firm. But it's hard to tell if a firm is treating its employees well or if it has a healthy culture from looking at the financial results. The financials can tell you how well the firm has done historically, but not whether it is making the right investments right now. If all you do is rely on the financial statements, there's an enormous amount of information you just don't see. And if you don't see it and can't measure it, you won't know that it exists or whether it matters.

So-called Environmental, Social, and Governance (ESG) metrics are one possible solution to this problem.[14] They have their origin in the 1980s, when a number of high-profile disasters, including the 1984 leak of toxic gas in Bhopal, India, that killed at least fifteen thousand people and injured many more and the 1990 Alaskan *Exxon Valdez* oil spill, led several NGOs to demand that firms disclose more information about the environmental and social effects of their operations.[15] In response several firms began issuing corporate social responsibility reports. These early reports contained only very limited amounts of quantitative information. Shell's 1998 offering, for example, consisted almost entirely of a chatty discussion of the firm's "General Business Principles."

In 1999, the Coalition for Environmentally Responsible Economics (CERES) founded the Global Reporting Initiative (GRI), an organization devoted to standardizing sustainability reporting.[16] The GRI issued its first set of guidelines in 2000, and by 2019, more than 80 percent of the world's 250 largest corporations used its standards to report on their sustainability performance, and its database had more than thirty-two thousand reports on file.[17] The GRI data are, however, of only limited use to investors. Their primary purpose is to highlight information that could help NGOs and governments hold corporations to account, and firms report the same information regardless of their industry, size, nationality, or ownership structure.

Since many investors suspect that better ESG metrics could help them generate superior returns, there has been an explosion of activity among entrepreneurs and nonprofits seeking to develop investor-friendly metrics. These efforts draw not only from GRI data but also from surveys sent to firms, annual reports, and a wide variety of public data. I know of at least two start-ups that are using artificial intelligence to construct information about social and environmental performance from scraping the web.

Despite the fact that many of these data are selectively disclosed, often difficult to compare, and of highly variable quality, they are already changing global investment practices.[18] More than 40 percent of all professionally managed assets—$47 trillion worth—are now invested using some form of social responsibility criteria.[19] Slightly less than half of this money is in so-called exclusionary funds—funds that exclude firms such as gun manufacturers or tobacco companies.[20] About 10 percent is managed through CEO engagement—active, hands-on work by investors that attempts to change firm behavior directly—while the rest is invested using

"ESG integration." In 2018, $19 trillion was invested this way, at least 20 percent of total assets under management.

Hundreds of studies have explored the degree to which performance against ESG criteria is correlated with financial performance. The results vary greatly depending on the metrics that are chosen and the structure of the study, but taken as a whole the evidence to date suggests that there is no relationship between these (very dirty) measures of ESG performance and financial success.[21] As a preliminary result, this is hugely encouraging, since it suggests that, at the very least, firms trying to do the right thing are not underperforming their competitors.

More recent work suggests that the way to move forward is to focus on the subset of ESG metrics that are *material*, that is, that capture aspects of nonfinancial performance that have a significant impact on profitability.[22] (Material events or information are any events or facts that would affect the judgment of an informed investor.)[23] Recent research using handcrafted data sets and metrics that are almost certainly material to the firm's economic performance has found convincing evidence that the two are positively correlated.[24] But developing these kinds of metrics has not been easy. Jean Rogers and her colleagues have spent nearly ten years bringing one set to life.

After graduating with a PhD in environmental engineering, Jean took a job with a company that cleaned up superfund sites.[25] She loathed it. She later said, "I hated that job because . . . it was just cleaning up messes, and it horrified me that people had let it get to this point and were OK with these end of pipe 'solutions' that did not solve the real problems." She moved to one of the big accounting firms, where she hoped to gain a business perspective on how to address environmental problems and then to Arup, a global professional services firm, where she became the leader of the firm's US management consulting practice.

Here she became radically dissatisfied with existing reporting standards. "I worked with many companies to develop sustainability reports, but they weren't being used as management tools," she later recalled. "The companies were only doing them so they could say they did them, using them as public relations. There was no comparability between industries, or even between companies within the same industry." Few industries had metrics. "The GRI relied on general indicators for most of their reporting; they had only defined industry-specific metrics for the five industries that had asked for them. I knew, and others agreed, that because different sustainability metrics were more or less important depending on the actions of companies in a given industry, reporting had to be industry-based, but when I brought it up in conversation, people said, 'Yes, but it's just too hard, there are too many industries and too many indicators.'"

In 2011 Jean—together with a number of other pioneers working at the intersection between sustainability and reporting—founded the Sustainable Accounting Standards Board (SASB, pronounced "sasbee").[26] Jean wanted to build a world in which any and all investors could type in a ticker symbol and bring up useful ESG data for any company as easily as they could bring up financial data. Her plan was to develop separate standards for each industry, so that companies need only report on the issues that were material to them. The data would be easily auditable and easily comparable across companies.[27] A focus on materiality would allow SASB to argue that every firm had a duty to report on precisely these metrics since they have a legal duty to report all material information. It would also mean that the metrics would be much more likely to be correlated with the company's performance—and thus be useful to investors. Jean believed that once these kinds of metrics had been defined and widely accepted, they would allow firms to communicate the value of strategic initiatives designed to create shared value more effectively, and that once investors were able to build a clearer sense of the link between ESG and financial performance, they would push the companies they owned to use them to improve the company's strategy.

Jean and her staff began by constructing "materiality maps" for each industry, searching tens of thousands of documents to develop an understanding of the issues that shaped performance in each of them and developing a preliminary set of metrics that were useful, cost-effective, comparable across firms, and potentially auditable. They then convened a set of industry working groups, drawing on investors, corporations, and other stakeholders, to produce a draft set of standards. Each draft was subject to further review by the full set of SASB's members, and was then made available for public comment for ninety days. By 2018 the group had released a full set of standards for seventy-seven industries.

Preliminary academic analysis, as I noted above, confirmed that the new standards were positively correlated with long-term financial performance.[28] Equally intriguingly, there was also evidence that they were helping firms attract investors with significantly longer time horizons. Consider, for example, Sophia Mendelsohn's experience at JetBlue.

Sophia joined JetBlue as its head of sustainability in 2011.[29] She had previously worked as head of sustainability in emerging markets for Haworth, a multinational manufacturer of furniture, and for the Jane Goodall Institute in Shanghai, where she established environmental programs in offices and schools across China.[30] When she joined JetBlue, the firm was throwing away one hundred million cans a year, and her first task was to put a recycling program in place. Then she began to think about what else she might do.

She focused first on creating shared value wherever she could find it. In 2013 she launched a resource efficiency program designed to (among other goals) decrease potable water use. The majority of JetBlue's flights were landing with their water tanks almost full, so

Sophia spearheaded the implementation of a policy that required potable water tanks to be only three-quarters full. The change resulted in a reduction of roughly 2,700 metric tons of CO_2 and fuel savings of nearly a million dollars a year.[31]

In 2017 Sophia and her team spearheaded a multifunctional initiative to introduce electric vehicles as ground service equipment at the JFK airport in New York, JetBlue's home base. The project was expected to cut operating expenses by roughly three million dollars over ten years, and had a net present value of nearly three-quarters of a million dollars. It triggered widespread interest across the industry, and at least one airport authority has used it as an example of how best to think the electrification of ground service equipment. Two years later, Sophia entered a binding agreement to purchase thirty-three million gallons of renewable blended jet fuel a year for at least ten years *at the same price as standard jet fuel.* It was the largest renewable jet fuel purchase agreement in aviation history and—since jet fuel was the largest single component of JetBlue's cost after wages and salaries, and oscillated wildly as a percentage of revenues—a major coup.[32]

Sophia's interest in accounting grew out of a conversation with her colleagues at JetBlue's investor relations department.[33] JetBlue had become one of the most profitable airlines in the United States by stressing a passionate commitment to customer service. This meant investing for the long term—not only in the kinds of technical investments that would immediately improve the customer experience (like multichannel TV), but also in building strong relationships with its crew members and in thinking through the ways in which its customers' focus on sustainability might shape their views of the airline. But despite this long-term orientation, many airline investors, including some of JetBlue's investors, were largely short-term-orientated. Sophia's colleagues in investor relations believed that if JetBlue could find a way to communicate its strategy more effectively, it would help to attract more long-term, growth-orientated investors. This, in turn, would make it easier to make the kinds of long-term investments most likely to fuel JetBlue's growth.

In response, Sophia proposed that JetBlue become the first airline to issue a SASB report. She argued that since SASB's airline metrics included both measures of the firm's relationship with its workforce and of its approach to sustainability, issuing such a report would be a powerful way to help communicate the firm's long-term, growth-orientated outlook, particularly since JetBlue was significantly ahead of its competitors on both dimensions.

In talking about the decision later, she said:

> Ultimately, what we want to do is . . . increase the value of our shares, diversify our investor base, (and) reduce volatility in the stock. So we want shareholders to believe in our stock for the long run. Our investors are our owners, and they deserve to have information in the way they want it—particularly as it relates to the major environmental and social mega trends pressuring the industry. Sustainability reporting has shifted from being about storytelling to being about model orientated data sharing.

This reporting strategy dramatically increased investor interest. One major investor spent two hours cross examining Sophia on the move. By 2017 turnover had grown from 30 percent in 2015 to 39 percent, the highest figure in the industry. Two years later an increasing number of investors were routinely asking questions that moved seamlessly from sustainability to the business. Sophia had also discovered that the process of putting together the report had powerfully reinforcing effects inside the organization, since framing sustainability as something that had a major effect on JetBlue's financial performance—a perspective that everyone understood—provided a way to talk about issues like climate change in a way that built commitment to the idea across the firm.[34]

Well-designed, material, auditable, replicable ESG metrics can thus play an important role in matching purpose-driven firms with investors who care about the long term and who view the creation of shared value as a route to superior profitability. They may also allow asset owners to solve some of the agency problems inherent in the fact that so many financial assets are professionally managed by those with relatively short time horizons, by making it significantly easier for asset owners to communicate a concern for the long term, and for social and economic performance, to the professionals who manage their money. Take, for example, the case of the Japanese Government Pension Investment Fund.

Hiro Mizuno joined the Japanese Government Pension Investment Fund (the GPIF) in the fall of 2014 as its chief investment officer.[35] He took a significant pay cut, leaving a high-profile private equity job in London to supervise eighty employees on a single floor of a rather ordinary office building in downtown Tokyo. Press coverage at the time noted that it was an unconventional decision, but no one suggested that it had the potential to spark a revolution in the way that one of the world's largest pools of money worked with its asset managers to address environmental, social, and governance issues.

The GPIF is the largest pension fund in the world, holding about ¥162 trillion (about US$1.6 trillion) in financial assets. Before 2013 the fund invested the majority of its portfolio in Japanese sovereign bonds, but in 2014 GPIF's regulators decided that the fund should diversify its portfolio and invest a significant fraction of its resources in equities (shares of publicly traded companies) in the hope of significantly increasing returns. This presented Hiro with a quandary.

There were two routes he could take to increase GPIF's performance. One was to try to pick winners by investing only in those firms that were likely to outperform their competitors. This approach makes intuitive sense and sometimes yields spectacular results. For example when Peter Lynch took over management of the Magellan mutual fund in 1977, it had only about $18 million under management. Lynch believed the secret to success was understanding individual companies in depth and investing in those he thought most likely to succeed.[36] He was spectacularly successful: between 1977 and 1990, the fund averaged a more than 29 percent annual return, making Magellan the best-performing mutual fund in the world.[37] By 1990 it had more than $14 billion under management.[38]

But Hiro knew that Lynch's story was an alluring exception, and that "active" investors—those like Lynch who try to invest only in high-performing firms—on average make consistently lower returns than "passive" investors who buy a defined group of equities and simply hold them.[39] Moreover GPIF is simply too big to be able to invest in only a limited set of firms. It owns about 7 percent of the Japanese equity market and roughly 1 percent of the world's, and it is also a huge investor in the bond markets. This means that the fund is what is known as a "universal investor"—an investor with so much money to invest that it is effectively forced to hold stock in every available firm.[40] Indeed 90 percent of GPIF's Japanese equity portfolio and 86 percent of its foreign equity portfolio are invested in "passive funds"—funds that hold every available stock in a particular class and that are designed to track the performance of the entire market.

Hiro therefore decided to try to improve GPIF's performance by improving the health of the entire economy—by persuading every firm in Japan (and indeed in the world) to embrace the use of ESG.

In Hiro's words,

> Private business is always built upon a competitive model. But GPIF is a public asset owner; we don't need to beat competitors or the market . . . GPIF is a super-long-term investor. We are a textbook definition of a universal owner. . . . Some people say ESG is

> not a positive attribute for achieving excess returns. But . . . we are not interested in making excess returns. We are more interested in making the whole system more viable.

Informally, he stressed the consistency of an approach to investment rooted in environmental and social performance with long-standing Japanese cultural values, remarking, "My grandmother would have been very upset if I told her that, in my position or my job, thinking about the global environment or social issues was against my professional mandate. She would tell me to quit my job immediately."

There were several reasons to believe that pushing firms to focus on environmental, social, and governance issues would improve the performance of the Japanese economy. Focusing on improving corporate governance—the "G" in ESG—seemed like the obvious place to begin. There is widespread agreement that one of the reasons Japanese firms have generated significantly lower returns than their foreign competitors over the last twenty years is that Japanese corporate boards are relatively weak by global standards. Many Japanese managers are so secure in their positions that they feel no pressure to exit underperforming businesses or to explore new opportunities. In 2017 only 27 percent of major Japanese companies had a board on which more than a third of the directors were independent, and many of these "independent" directors were lawyers or academics with little management experience. GPIF's first order of business was thus to try to persuade its asset managers to push the companies they owned to improve their governance structures so as to give more power to investors—to ask them to disclose more information about their businesses, to talk to their shareholders about long-term strategy, and to vote their shares with governance in mind.

Focusing on social issues, or on the "S" in ESG, also seemed likely to yield substantial dividends. Japan's birth rate had dipped below replacement levels in the mid-1970s, and Japan's working-age population was declining faster than any other on the planet.[41] Given Japan's closed immigration policies, persuading more women to stay in the labor force was critically important to long-term economic growth. But making this happen required dealing with some deep-rooted structural problems. Many Japanese companies have a two-track employment system. New employees are sorted into the *sogoshoku* (managerial) or the *ippanshoku* (general clerical) track. Participation in the *sogoshoku* track is critical to securing regular employment status and the possibility of promotion to management positions, but women are disproportionately hired into the *ippanshoku* track. Women are also expected to take primary responsibility for child-rearing, and since most employers expected their employees to work very long hours, it is difficult to combine having children with a career. In the World Economic Forum's 2017 Global Gender Gap Index, Japan ranked 114th out of 144 countries.[42]

Hiro also believed that attempting to solve Japan's environmental problems—the "E" in ESG—was critical to ensuring the long-term well-being of his beneficiaries. Unchecked climate change threatened to destabilize Japan's food supply and to fuel an increase in the frequency of natural disasters in an island already disproportionately subject to them. "Even if I could pay pensions thirty years from now," Hiro said, "what good would it do if the grandchildren of my beneficiaries can't play outside?"

Deciding to focus on ESG was one thing. Implementing the decision was another. As an independent administrative agency, GPIF was prohibited from directly trading equities or from directly talking to firms to minimize any potential government influence over the private sector. All of the fund's investing was outsourced to independent asset managers.[43] Hiro thus began by asking every one of GPIF's thirty-four asset managers to start a systematic conversation with every company they invested in about that company's approach to ESG issues, to vote in each company's proxy election, and to report that vote to GPIF. For example, an asset manager might note that a particular board's nomination and governance committee had no independent chair, ask when that would be rectified, and threaten to vote against the management if it was not. GPIF would meet one-on-one with each asset manager at least twice a year, asking each of them to outline how they were engaging with companies, and requiring them to disclose their proxy votes.

To say that all hell broke loose—in a very Japanese way—is to understate the case. One observer suggested that Hiro's suggestion that he would like GPIF's passive managers to be "passively active" was "the most controversial announcement in the history of asset management." Nearly all of the fund's asset managers protested—quietly and politely—that they didn't have the expertise required to make informed decisions with respect to ESG. The active managers objected that their expertise lay in increasing alpha (risk and market-adjusted relative return) and that they were unconvinced that focusing on ESG would help. The passive managers suggested that since they were paid less than 0.1 percent of an investment's value, they couldn't afford to develop the necessary capabilities.[44]

Hiro replied—equally quietly and politely—that it was not his intention to reduce anyone's compensation and that he was happy to pay for better performance and increased expertise. He also pointed out—very diplomatically—that none of the asset managers should feel forced to work for GPIF. He then changed the way in which GPIF evaluated, selected, and compensated its asset managers, signing multiyear contracts with the active asset managers who were willing to agree to a new fee structure, designed to reward them explicitly for generating alpha and for focusing on the long term, and asking his passive managers to propose a "new business model" for their fees. When, after two years, none of the passive managers had made a concrete proposal, he increased the weight GPIF placed on stewardship activities in the selection criteria to 30 percent and announced that asset managers who did not meet the new expectations could see their investment allocation reduced or, at worst, lose GPIF's business. He also reiterated his offer to change the fee structure.

Hiro's investment consultants were uncomfortable with the new contracts, suggesting that giving asset managers multiyear contracts violated GPIF's fiduciary duty since the multiyear commitment sacrificed the option value inherent in being able to fire managers at any time. Hiro replied that it was failing to sign long-term contracts that violated GPIF's fiduciary duty since it promoted short-termism. He said, "They did not understand my fiduciary duty correctly. I have a fiduciary duty across generations."

Hiro also used GPIF's size and visibility to raise the profile of ESG issues across the entire Japanese business community. He launched five stock indexes based on ESG themes, and invested about 4 percent of the money that GPIF had allocated to equities, about ¥3.5 trillion (about US$32 billion), into them.[45] He also ensured that the methodologies that had been used to construct each index were fully disclosed. This was not a usual practice. Hiro noted the following:

> If I think more conventionally . . . and define my job as beating the market, we shouldn't have asked the [index] vendors to disclose their methodologies. Nevertheless, we demanded that they disclose them, because what we wanted was to improve the whole market, not to beat the market. With access to the selection criteria, companies that were not selected for the indices could learn how to improve their ESG rating. And we demanded that the index vendors engage with companies and report their progress. The vendors say that the number of inquiries they have received from Japan has drastically increased since the debut of the indices.

Hiro has changed how Japanese investors and Japanese companies think about the potential to create shared value and the importance of investing in social and environmental issues. Press mentions of ESG have increased more than eightfold between 2015 and 2018.[46] Among large and midsize companies, 80 and 60 percent, respectively, say that the increasing focus on ESG indexes has raised awareness about ESG at their companies and has led to real change,[47] and nearly half of Japanese retail investors claim to recognize the importance of considering ESG in their investment decisions. In the last two years the percentage of Japanese financial assets allocated to sustainable investments has increased from 3 to nearly 20 percent.[48]

Clearly much remains to be done, but Hiro's preliminary success is deeply encouraging. The widespread use of material, replicable, comparable ESG metrics is a game changer, potentially enabling investors to develop a much richer understanding of the relationship between a firm's investments in social and environmental performance and returns to the individual firm—as at JetBlue—and returns to the portfolio as a whole—as at GPIF. But the adoption of ESG metrics alone is clearly not currently sufficient to fix the short-termism problem. Many ESG metrics remain hard to construct, rarely comparable across firms and often difficult to audit, and even those that are well thought through are insufficient to capture the universe of useful nonfinancial factors that may drive performance.

Developing widely adopted, standardized metrics that can be routinely incorporated into financial statements will take some time. Moreover, even with the best ESG metrics in the world, it will be difficult to convincingly communicate the value of some of the more intangible investments that enable purpose-driven firms to succeed. As my description of SASB suggests, this is an area of hugely active research, and things may well change going forward.

In the meantime, though, it's worth exploring some of the other solutions that have emerged to focus capital on the long term.

One possibility is to move away from the public capital markets altogether. Family-owned firms, for example, are in principle ideally positioned to focus on long-term value creation, and family-owned firms like Tata and Mars are among the world's most purpose-driven firms. But while there is some evidence that this is indeed the case, the performance of family-owned firms tends to be highly variable, and indeed many development economists believe that a need to rely on family ownership is one of the factors that depresses economic growth in those countries that have yet to develop widely trusted public capital markets.[49]

Private equity funds offer another source of highly informed, long-term capital, but again the evidence on this front is mixed. Private equity funds appear to outperform the public markets, but so far as I am aware, there is no systematic evidence that they are more focused on the long term than the public capital markets.[50]

Another possibility is to look to investors who are themselves purpose driven and who share the firm's goals and its commitment to the long term. The bad news is that there are not very many of them. The good news is that this is starting to change, and that the firms they invest in are more than capable of holding their own in the face of more conventional competitors.

Find Investors Who Share Your Goals

So-called impact investors are the financial equivalent of purpose-driven firms like Unilever or King Arthur's Flour. They seek a decent return, but their goal is to make a difference in the world rather than to maximize profits. It's a group that includes not only arms of philanthropic foundations like the Bill & Melinda Gates Foundation and the Omidyar Network but also wealthy individuals and families, private equity firms, and even a few institutional investors.

Reynir Indhal, the private equity partner who funded the purchase of Norsk Gjenvinning and was one of Erik Osmundsen's strongest supporters in his efforts to turn around the firm, now leads Summa Equity, a private equity fund whose website proudly proclaims "We Invest to Solve the World's Global Challenges."

Triodos Bank is a particularly visible example of the power that these kinds of institutions can exercise—and also of the kind of commitment that is required to build them.[51] The bank, which is based in the Netherlands, began as a four-person study group devoted to exploring how money could be managed more consciously. The participants were interested in a system of spiritual philosophy developed by the philosopher-scientist Rudolf Steiner, who conceived of society as comprising three realms: economic, rights (including politics and law), and cultural-spiritual. He believed that a healthy society depended on a balance among the three domains. The founders decided that the purpose of their new endeavor would be to initiate social change by stimulating innovative entrepreneurial activity. Triodos Bank incorporated in 1980, equipped with €540,000 in start-up capital and a banking license from the Dutch Central Bank.[52] The bank is owned by its customers, a strategy that has allowed it to pursue goals that are explicitly about building a healthy society, rather than about maximizing short-term returns. Today it has more than €15 billion in assets under management and €266 million in revenues.[53]

The word *triodos* translates as "threefold way," and the idea that the bank would support the healthy development of Steiner's three societal domains has always been integral to its purpose. Eric Holterhues, the head of arts and culture within Triodos Bank Investment Management, described the mission-centric purpose of the bank in this way:

> Three things are important for society. Preserving the Earth—that's why we're active in all kinds of environmental projects. Then, how

> people cope with each other on the Earth—that's why we're active in fair trade, in microfinance. And finally the development of each individual—that's why we're active in culture. This also makes us different from other banks. We don't say, "We are a bank, and let's see what the sectors are where we can make money." We say "These three sectors: the Earth (environment), we (social) and I (culture), this is our starting point and how can we contribute to them as a bank?"

Peter Blom, the bank's CEO, described the bank's vision like this:

> You want to influence what's happening in the next 10 years. . . . That is already a difference from many other banks. Maybe they think about the future a little bit, but it's more like "How can we do things we already do better?" Not so much, "What do we want to influence and change?"
>
> That's a very important notion and if you do that you need to think about big trends in society. Where are we going? Where is mankind going? What is essential for people? Then you want to be able to look back after ten to fifty years and see what has worked out. That's the sort of approach where you learn from the future, going back to where we are now. If you don't do that, then it's very easy to repeat and repeat. You have to understand the spirit of the time. The spirit of the time is connected to the longer development of people—of entrepreneurs and how we will do business.

Finding measures that might capture these kinds of goals is not an easy thing, and Triodos Bank is a living example of the limitations of a simple reliance on ESG metrics, however sophisticated. Loan decisions, for example, involve a subtle process of individual and collective judgment, requiring the loan officer to work out whether an application is aligned with the bank's mission *and* presents the right risk profile. One phrase that is often used around the bank is, "If a child came to ask you for five euros, your first question would be 'Why do you want it?'" As Pierre Aeby, the chief financial officer, described it, "When we invest in a loan, we look first at what the loan is for. What is the mission of the borrower? What value does the borrower add in society? What is the match with our own values? Then we look at it strictly as a banker. What is the repayment capacity? What do they do? What is the collateral? Then we fix the market price."

Daniël Povel, who managed business lending in the Netherlands for the bank, explained that "It's very easy to finance, say, an organic/biodynamic farmer who hires former drug addicts to work the farm, who also runs an art center that produces paintings and sculpture, and who wants a loan to put solar panels on his roof. That's very easy; everyone would say yes." But projects often fell into a grey zone. The bank described these as "dilemmas" that required individual judgment and discernment in dialogue with one's peers to make a decision. Daniël told this story of a dilemma that was discussed at one of the bank's weekly Monday morning meetings:

> We were asked for a loan by a shoe factory, a very famous one, at least in Europe. They were trying to cut their energy costs by being more energy efficient. They wanted a loan to make their own energy with the leather bio-waste from the factory—a fatty, oily substance that they collect when they scrape cow skins. They wanted to burn this substance to generate heat and electricity in order to cut their energy use by 30 or 35 percent, which is a lot. They were advanced on other environmental issues, for instance a lot of chemicals are used in shoe production, and they purified the water so it could be used almost as drinking water.
>
> My colleague who presented this example at the meeting asked, "Are we going to finance it?" And someone asked, "These cows that provide the skin for the shoes, are they allowed to walk outside?"

> And he responded, "Do you want to have barbed wire marks on your shoes? Of course they don't walk outside. They are contained inside." The problem with this is that Triodos Bank doesn't want to support intensive animal farming operations that confine animals to overcrowded indoor spaces and never allows them the freedom to move around or go outdoors. So the question is, what should we do?[54]

In short, Triodos Bank relies on guidelines rather than on quantitative criteria. People are asked to think about projects using not only their head but also their heart and their gut. As Blom phrased it, "We deliberately never use the word 'criteria.' We've always called them 'guidelines,' and that creates room for discussion and dialogue. That is something more situational, more living and probably involves much more explaining than having abstract criteria."

Triodos Bank generates financial returns in the 5–7 percent range—significantly below those earned by the large global banks in their best years,[55] but significantly above those earned in their worst years. It has also succeeded in making a significant difference in the world. It launched Europe's first green fund, the Biogrond Beleggingsfonds (the Ecological Land Fund), dedicated to funding environmentally sustainable projects in Europe. A Wind Fund soon followed. At the time, wind energy technology was in its early stages, but the bank identified promising wind turbine manufacturers in Denmark, Germany, and the Netherlands, and a small engineering company in the Netherlands as a potential partner. The fund was profitable from the beginning, and had a significant impact on the wider industry as other banks began to offer similar products. Blom summed up this approach to system-level intervention this way:

> You take a strategic view. You know the sector, and you say well this is missing and in a healthy sector, those elements are needed,

> and there will also be demand from customers of course. How do we now find the entrepreneurs, who want to take the next step with us?
>
> We're trying to initiate a virtuous circle, where the loans we make act as demonstration projects and there are spillover effects—so other people come into the space. We're looking for trust—so we try not to be a typical bank that's always out to make the best deal for itself. Clients come to us because we are a hub. We're building a whole new set of skills in support of a much more collaborative economy.

In summary, Triodos Bank is pursuing a classically purpose-driven strategy, using its focus on the broader community to catalyze the kind of architectural innovation that can change the whole system. The sixty-four thousand dollar question is whether these kinds of investors—investors who value the well-being of the planet over squeezing the maximum possible returns from their money—are a fringe group operating at the margins or the wave of the future. I don't think we yet know the answer. But I am hopeful. About $68 trillion of wealth is expected to change hands in the next twenty-five years as the baby boomers die, and much of this wealth will be given to a younger generation that is much more interested in impact investing than their parents were.[56]

Another route to securing investors committed to the long term is to raise capital from customers or employees. Triodos Bank's success, for example, is critically dependent on the fact that it is owned by its customers—customers who choose to own the bank because they share its mission and values and are committed to its long-term success. KAF's CEOs believe that the fact that firm is entirely employee-owned makes possible a degree of engagement and a commitment to the mission of the firm that would not otherwise be possible.

Customer-owned firms are surprisingly widespread. Rural electricity cooperatives played a crucial role in electrifying the United

States and still provide electrical power to more than 10 percent of the US population.[57] Many farmers who found themselves at the mercy of concentrated buyers have responded by creating farmer-owned cooperatives, such as Land O'Lakes or Dairy Farmers of America.[58] These groups aggregate the collective power of individual farmers to ensure that they receive the market price, and often support marketing campaigns in support of sales. There are currently about four thousand customer-owned agricultural cooperatives in the United States, with roughly $120 billion in revenue among them.[59]

The early years of the insurance industry were dominated by customer-owned "mutual" insurance companies, as investor-owned insurance companies initially focused on charging the highest possible premiums, while customer-owned companies were much more likely to write policies that rewarded risk-reducing behavior.[60] There are still about twenty thousand mutual insurance companies in the United States. Credit unions are customer-owned cooperatives founded to provide the best possible service to their members without the constraint of having to maximize investor returns. There are currently approximately fifty thousand of them in the United States. Collectively, the credit unions and the mutual insurance companies have about $180 billion in revenues and employ more than 350,000 people.[61]

These are relatively small numbers by the standards of the world's biggest agricultural buyers or the world's biggest banks. The world's two biggest traders in agricultural commodities, for example, have more revenue between them than all four thousand agricultural cooperatives.[62] The two largest American banks have more revenue between them than all fifty thousand financial cooperatives.[63] But their existence suggests that customer ownership might be a crucial piece of a reimagined capitalism.

Employee ownership is relatively common, although in most cases employees have essentially no control over the management of the firm. In 2013, about 38 percent of American employers offered profit sharing;[64] 20 percent of employees reported owning stock in their employer;[65] about 5 percent of employees took part in Employee Stock Option Plans (ESOPs); and about 15 percent took part in Employee Stock Purchase Plans (ESPPs), which allow employees to buy stock in their employer. Participation in such plans was most common for managers or salespeople, but employees of all types could and did participate.[66] In some of these companies employees owned a majority (or a large minority) of the company. For example, employees owned Avis, the car rental service, until they sold it to an outside investor in 1996. In 1994, United Airlines' employees agreed to an ESOP, acquiring 55 percent of company stock in exchange for salary concessions, making the carrier the largest employee-owned corporation in the world at the time. Employee ownership of United ended in 2000.

Employee control is much less common, but it is attracting an increasing amount of interest, particularly as a solution to the problem of inequality. Michael Peck, Mondragon's delegate to the United States, is the executive director of 1worker1vote, a nonprofit designed to support the formation of cooperatives modeled on Mondragon. The organization is supporting work in ten cities across the country and has forged alliances with a wide range of organizations, including the United Steel Workers, the National Cooperative Bank, and the American Sustainable Business Council.[67] In Preston, England, the local council is actively experimenting with worker-owned cooperatives as a step toward the revitalization of the city.[68]

In the United States the Publix supermarket chain is the largest employee-controlled firm with more than one thousand locations throughout the Southeast and two hundred thousand employees.[69] The largest employee-controlled firm in the United Kingdom is the John Lewis Partnership, which operates about forty department stores and three hundred grocery stores throughout the country.

In 2017 it had revenues of just over £10.0/$12.3 billion. It is a public company whose stock is owned in trust for its eighty-three-thousand-plus employees (known as "partners" within the firm). The firm is governed by a partnership council whose members are elected every three years by a vote of the partners; a partnership board, which is elected by the council and serves as the board of directors; and a chairman, selected by the board.[70]

Mondragon, a firm based in Spain's Basque region, is the largest employee-owned firm in the world.[71] In 2018 it had €12.0/$13.2 billion in revenues and employed more than eighty thousand people.[72] It is run as a cooperative, so that worker-owners participate directly in the management of the company on a one worker–one vote basis, and ownership is nontransferable. When workers retire or leave, they receive a financial package or a pension in exchange for their ownership stake, rather than being able to sell their shares. Mondragon is a holding organization for more than one hundred worker cooperatives. Together they compete in dozens of industries, including heavy manufacturing (auto parts, home appliances, industrial machines); light manufacturing (exercise equipment, antique firearms, furniture); construction and building materials; semiconductors; information technology products; business services (human resources management, consulting, law); education; banking; and agribusiness. Mondragon invests heavily in education (Mondragon University is a nonprofit cooperative with some four thousand students), and runs its own bank and consulting company—all dedicated to helping its member cooperatives to succeed and to spawn new ones. In 2013 it won one of the *Financial Times*' "Boldness in Business" awards "for what it represents in terms of a real proposal for a new type of business model: 'Humanity at work,' based on co-operation, working together, solidarity, and involving people in the work environment."[73]

Employee-owned firms—as you might expect—appear to prioritize employment over profits and to pay their employees above the going wage.[74] One study found that 3 percent of employee-owners were laid off in 2009–2010 compared to 12 percent for nonemployee owners, and that employee-owners had approximately more than twice the amount in their defined contribution accounts as participants in comparable nonemployee-owned companies, and 20 percent more assets overall.[75] One senior manager at Mondragon suggested that the firm played a significant role in reducing inequality, claiming that "if the Basque region in Spain were a country, it would have the second-lowest income inequality in the world." Employee-owned firms also grow faster and are vastly overrepresented in the "Best Companies to Work For" rankings.

When employee ownership is linked to the ability to participate in decision making and is accompanied by greater job security, it increases employee loyalty and motivation, lowers turnover, and drives higher levels of innovation and productivity.[76] The fact that King Arthur Flour is owned by its employees, for example, makes it much easier to make the investments that are essential to building and maintaining a highly engaged workforce—not only to providing training, decent wages, and benefits, but also to invest the time and energy required to ensure that information is widely shared, that the culture is sustained, and that everyone is engaged.

Both customer and employee ownership thus offer potentially promising pathways forward to rewiring finance, and building their share and presence in the economy is likely to be a critical building block of a reimagined capitalism. Leveling the legal and regulatory playing field to make employee- or customer-owned firms easier to create is an important policy goal for those interested in building a more equitable and sustainable world. But at the moment they are more a promising model for the future than an immediate solution—a promising project for purpose-driven millennials!

Perhaps this is why so many people working in this space have decided that the only way to force investors to focus on the long term is to change the rules of the game so that they have less power over the firm.

Change the Rules of the Game to Reduce Investor Power

Many of the people I most respect believe if we are to build a just and sustainable world, we need to reject the idea of shareholder primacy altogether.[77] They believe that the only way to construct a sustainable capitalism is to adopt a different view of the firm—one in which managers and directors owe their loyalty not to investors but to the "stakeholders" in the firm—to investors, yes, but also to employees, suppliers, customers, and the community itself. They are keen to change the legal rules that control how firms are governed to make this a reality.[78]

I am deeply sympathetic to the idea that we need to reduce the power of investors in some important ways. But I think the path to doing this is both more complicated—and potentially less straightforwardly beneficial—than some of its proponents suggest.

I am, for example, a fan of the legal form known as the "benefit corporation."[79] Firms that incorporate as benefit corporations formally commit to creating public benefit as well as to giving decent returns to their investors. The company must publish a strategy outlining just how it is planning to do this,[80] and the board has formal responsibility for making decisions that create public as well as private value. Benefit corporations must also produce an auditable report every year detailing their progress toward generating the public benefit they have promised to create. You can incorporate as a benefit corporation in thirty-six US states[81]—including Delaware, and there are at least 3,500 benefit corporations in operation, including Kickstarter, Patagonia, Danone, Eileen Fisher, and Seventh Generation.[82]

Choosing to incorporate as a benefit corporation offers a number of tangible advantages to firms hoping to make the world a better place. It makes clear that neither the directors nor the managers have a legal responsibility to maximize shareholder value. Indeed directors are required to consider the public interest in making every decision. Most importantly, when directors have committed to sell the firm, they can select the buyer that will create the most value for all the firm's stakeholders, rather than the one that offers current shareholders the most cash. This is hugely important. As noted earlier, directors at conventional firms do not generally have a legal duty to maximize shareholder value unless they have made a decision to sell the firm, and current shareholders will not have voting rights in the new entity. If this is the case, US directors have a legal responsibility to sell the firm to the buyer that offers the highest price.[83] This may seem like a detail, but it is not. In a conventional firm, the fact that there is always a risk that the directors may be forced to sell the firm to the highest bidder can make it much harder to make the kind of long-term investments—in building trust, in treating one's employees well—that are essential to building "high commitment firms." Firms that are subject to the whims of the financial market are untrustworthy partners, which can make it much more difficult for them to build the long-term, trust-based relationships that are so essential to building purpose-driven firms.[84]

Does this mean that making all firms benefit corporations is the secret to reimagining capitalism? Alas, probably not. Reducing the power of investors is a double-edged sword. If the firm's investors, the board of directors, and the management team are firmly committed to doing the right thing—and will bend every nerve and sinew to make it happen—then incorporating as a benefit corporation makes enormous sense. The firm's managers won't have to rely on necessarily imperfect measures of shared value creation to convince investors to let them move forward, and to the degree that focusing on public value creation increases profitability, investors may even be better off. What's not to like?

There are two problems. The first is that the model is heavily dependent on the firm's ability to attract investors who share the mission of the firm—or who believe that operating this way is a reliable route to increasing profitability. In a benefit corporation all the power remains with the investors. Only they can elect the directors. Only they can sue to enforce adherence to the mission.[85] In the worst case, ruthless investors can take control of the company by voting in a new board, give only lip service to the creation of public benefit, and simply re-create a conventional firm.

The second problem is that—alas—not all managers and boards can be trusted. Unless and until ESG metrics are sufficiently well developed that they can allow investors to determine with a fair degree of certainty whether a firm is creating public benefit, both managers and boards will have the temptation to use the benefit corporation structure as a way to take life easy. Of course, if investors are willing to sue, and the firm's metrics are sufficiently detailed and closely linked to performance, this won't be a problem. But it is a strategy that puts an enormous premium on good metrics and on deep engagement between investors and the management team.

In Japan, for example, the "miracle" that remade Japan following World War II stressed lifetime employment, close relationships with suppliers, persistent investment over long time frames, and an almost obsessive focus on the customer.[86] The approach was complemented by tight relationships between Japanese firms and their investors. Japanese firms had historically raised the bulk of their capital from banks, and in most firms the board of directors was staffed exclusively by company insiders and chaired by the CEO. While many firms were publicly listed, they were protected from the threat of takeover by a system of extensive cross-holdings.[87] In practice, Japanese managers could do almost anything they wanted without threat of pushback from their investors.[88]

This approach worked fantastically well—until it didn't. Between 1960 and 1995, it enabled Japanese firms to build purpose-driven, customer-obsessed firms like Toyota, and to conquer the world with innovative, low-cost products of unsurpassed quality. Japan's GDP grew at an extraordinary rate. In 1960 Japan's GDP was just over 60 percent of the United Kingdom's. By 1995 it was *four times* the size.[89]

But beginning in 1995, the Japanese economy flatlined. Between 1995 and 2017, the UK economy approximately doubled in size. But over the same period, the Japanese economy barely budged.[90] It is still the fourth-largest economy in the world—roughly the size of the British and French economies combined—but Japanese rates of productivity growth are roughly half of those in the United States and Europe, and the economy has been essentially stagnant for twenty years, a period that has become known variously as "the lost decade" or the "lost twenty years."[91] The question of just what caused this slowdown remains hotly contested, with explanations ranging from a growing demographic crisis and the toleration of gross inefficiency in some highly protected sectors of the economy to the combination of a massive asset bubble and a failure to hold Japanese banks accountable for its consequences. But many Japanese observers believe it also reflects the failure of Japan's system of corporate governance—that the very features that enabled Japanese firms to focus on the long term with such success in the sixties, seventies, and eighties are now a major liability. Japanese managers remain firmly in control at most Japanese companies, and as a result Japanese firms are relatively slow to exit underperforming businesses and/or explore new opportunities.[92]

Giving managers significant control over their firms is a high-variance bet. If they are competent and trustworthy, it gives them unparalleled freedom to make the kind of hard decisions that build great firms. If they are deeply embedded in a network of institutions that effectively holds them accountable—as was the case in the United States in the fifties and sixties, and has often been the case in countries such as Germany and the Netherlands, stakeholder-orientated governance systems can be very effective. But if these institutions change in fundamental ways, managers who have learned not to fear their investors may become entrenched opponents of change.

This is not just a Japanese issue. In the last fifteen years, many of the most successful Silicon Valley firms have gone public with dual class stock that has left founders in sole control of their firms. Facebook, for example, issued two classes of shares when it went public. The Class A shares went to everyday investors and came with one vote per share. But the founders—mostly Mark Zuckerberg—got Class B shares. Every Class B share came with ten votes. This means that it is effectively impossible to force Zuckerberg to step down, no matter how badly Facebook performs.[93]

In general, the founders in question claim that this structure is necessary to protect them from shareholder pressure. One observer, commenting on Snapchat's decision to leave the vast majority of voting power in the hands of its founders and to issue public stock with no voting rights at all, claimed that shareholders pressuring companies to cut expenses and increase short-term profits could stymie founder-led technology companies from making important long-term investments in major value-creating innovations. He suggested that "Great innovators simply see things that mere mortals cannot. As such, they are often out of sync with the wisdom of the crowds."[94]

But sometimes "great innovators" are simply once great founders who have run out of road and now refuse to see that the company needs to move in new directions. My point is not that changing the rules might not be a good thing—indeed if I were in charge, I'd require every publicly traded firm to change its governance structures such that it is no longer under constant threat of having to sell the firm for the highest possible figure—but that simply changing the rules is not an automatic or a costless solution to the problem of short-termism.

THERE ARE—BROADLY—THREE ROUTES to rewiring finance. One is to reform accounting so that firms routinely report material, replicable, auditable ESG data in addition to financial data. The widespread adoption of standardized, easily comparable, auditable ESG metrics would make it easier for firms to attract investors who will support them in making the kind of long-term investments that are essential to building successful purpose-driven firms and to creating shared value. More broadly, the right kind of ESG metrics could provide a richer language for talking about how doing the right thing can generate financial returns. They might lengthen time horizons and help both managers and investors unpack the dynamics of the relationship between doing good and doing well. When does it make sense to invest in human capital? To have a leading-edge environmental strategy? To clean up one's supply chain? As answers to these questions emerge, lagging firms will be pushed to catch up to the pioneers. It would be a different world. A world in which investors routinely insisted that firms invest in energy conservation. A world in which many employees were routinely better paid and better treated.

A second option is to rely on impact investors—or on one's employees or customers—for funding. This is a solution with many strengths, but it may be challenging to take it to scale. The third is to change the rules that govern corporations to shelter managers from investor pressure. This is intuitively appealing but would have to be managed with care. There's also the potentially significant problem that—at present—the vast majority of the world's existing investors would almost certainly fight the idea tooth and nail.

Rewiring finance will make an enormous difference and has the potential to support thousands of firms in their attempts to solve the big problems at scale. Can these kinds of investments make a difference in the grand scheme of things? It depends, of course, but there are several pathways through which individual firm action can have a significant effect on the big problems. Very large firms have a measurable effect simply through their own actions. Walmart works with nearly three thousand suppliers, who in turn work with thousands more.[95] Nike and Unilever similarly touch thousands of suppliers and millions of consumers. To the degree that they insist on better treatment of the employees or better environmental practices, they impact millions of people. But even much smaller firms change lives.

The pursuit of shared value can also have significant impact through its effects on other firms. Sometimes the simple demonstration that a particular investment makes commercial sense can persuade everyone in the industry to adopt the same practice. When Lipton demonstrated that sustainably grown tea cost only 5 percent more and that consumers cared sufficiently about the issue to increase Lipton's share, all of the firm's major competitors embraced sustainability too. Walmart's massive investments in energy saving and waste reduction have helped to persuade many other firms that these investments are likely to yield rich returns.

Successful purpose-driven firms can also shape consumer behavior. Twenty years ago, for example, most consumers assumed that "sustainable" meant "compromised"—that sustainable products were by definition more expensive or lower quality. That perception has steadily shifted as more high-quality products have come to market proudly flaunting their sustainability credentials. This, in turn, is persuading an increasing number of consumers to assume that it's possible to create fabulous sustainable products, and to demand this of more of the products that they buy. Leading-edge firms can also shape the cultural conversation. As the Nike case suggests, for many years no one held individual firms responsible for the behavior of their suppliers. Once that changed, the pressure to raise the bar substantially increased, and now nearly every major firm pays at least lip service to conditions in its supply chain.

Individual firms can also move the technological frontier. This is most evident in renewable energy, where every firm that enters the industry helps to drive down costs. Between 2015 and 2018, for example, Tesla installed a gigawatt-hour of energy storage technology (for comparison, in 2018 the entire world installed only slightly more). Since 2010 Tesla's efforts have helped to drop the price of battery storage by at least 73 percent.[96] New farming technologies introduced by firms like Jain Irrigation and John Deere are rapidly becoming industry standard, making it cost effective for many farmers to use water and fertilizer much more efficiently.[97] Sometimes the innovation is not technological, per se. Solar City, for example, pioneered a new model of financing solar panels that greatly expanded demand and saw the idea spread across the industry.[98]

Firms can thus help to kick-start a number of reinforcing processes that have the potential to drive change at scale. By demonstrating a new business model—and in the process potentially driving down costs and persuading consumers to demand it—they can push competitors to adopt the same practice, diffusing it widely across the industry. This process is sufficiently well-advanced in the food business that it is beginning to change global agricultural practices, and in energy it may be strong enough to play a significant role in driving the transition to fossil fuel–free energy. It's well underway in the construction business, and by some reports over half of all new construction in the United States is now built to energy-efficient standards.

But action by individual firms is a route to change that is inherently limited. In the end firms must be able to see a path to profit if they are to invest resources at scale, and this is much easier in some kinds of industries and with respect to some kinds of problems. To date, the big opportunities appear to be in industries and places where environmental degradation presents a clear and present danger to ongoing operations or to long-term sources of supply. Nearly all the world's major agricultural producers and traders, for example, are at the very least aware that they should be thinking hard about these issues, and many are doing much more. Using resources more efficiently also appears to be another significant opportunity. For years energy and water were so cheap that no one paid much attention to them. That is changing. To the degree that it is indeed the case that treating employees better improves their performance, there will be a business case for addressing inequality. Consumer preferences may shift dramatically, and in industries like food, consumer goods, fashion, and perhaps transportation, becoming more sustainable may increasingly be seen as a potential route to profit.

But this list still leaves plenty of problems that can't be addressed by firms working alone. Some problems are simply too large for any single firm to be able to build a business case around them.

The world is rapidly running through the available stock of wild fish—but every individual fisher has strong incentives to keep fishing if no one else is holding back. The falling price of renewable energy means that an increasing fraction of new power plants will be built using solar or wind. But heading off the worst consequences of global warming will require decommissioning many existing fossil fuel plants—and that's a process that without a change in the rules is very unlikely to be profitable. Building an engaged, well-paid workforce can be a potent source of competitive advantage—but it can be hard to pay people well and to treat them decently when one's competitors are busy racing to the bottom. Many firms would like to see the quality of local education improve, but very few can build a case for being the only firm willing to invest in making it happen. Many firms would like to see an end to corruption or an increase in the quality of local legal institutions, but most cannot make progress against either goal on their own.

Hiro himself faces a variant of this issue. He believes that averting the worst effects of climate change is very much in the best interest of his beneficiaries. But he faces several free-rider problems in trying to change behavior. The first is that it may not be profitable for individual firms to reduce fossil fuel use. Should Hiro instruct his asset managers to force them to? The second is that even if he has the power to force Japanese firms to go green, it seems wildly unlikely that he has the power to force every firm in the world to change. And if he can only change Japan, and global warming proceeds anyway, has he really done the right thing for his beneficiaries?

In short, many of the problems we face are genuinely public goods problems—and can only be solved through cooperative action or government policy. Is such action possible? Can firms and/or investors come together to solve the world's big problems? The next chapter explores what's happening on this front and asks whether and under what conditions cooperative action within industries or regions might help us reimagine capitalism.

6

BETWEEN A ROCK AND A HARD PLACE

Learning to Cooperate

> You can't stay in your own corner of the forest waiting for others to come to you. You have to go to them sometimes.
>
> —A. A. MILNE, *WINNIE-THE-POOH: THE COMPLETE COLLECTION OF STORIES AND POEMS*

Is rewiring finance enough to reimagine capitalism? Alas not. If every firm on the planet adopted a purpose beyond profit, pursued a shared value strategy, and was supported by sophisticated investors committed to the long term, it would be an enormous step forward, but nowhere near enough to solve huge problems such as climate change and inequality. Too many of these problems are genuinely public goods problems—solving them would benefit everyone, but no single firm has the ability to solve them on its own. We won't solve the climate crisis, for example, until and unless we can agree to leave the great forests standing. But if your competitors won't stop cutting down the trees, you, too, may have to cut them down to survive. We won't solve inequality until we spend more on education. But if your competitors won't train their employees, you won't be able to afford to train yours either. This is the tension we face. On one side the rock—the knowledge that continued deforestation and accelerating inequality may cause enormous harm—and on the other the hard place—every individual firm's inability to do anything about it on their own.

Industry-wide cooperation, or as it's sometimes known, industry "self-regulation" is one potential solution. This isn't an entirely crazy idea. Elinor Ostrom was awarded the Nobel Prize in Economic Sciences in 2009 for her work describing successful voluntary efforts within local communities to protect common resources such as forests and water. Her work suggested that local coordination could endure over generations and was often more effective than government action.

Many of the central institutions of the nineteenth-century American economy—including the New York Stock Exchange, the Chicago Board of Trade, and the New Orleans Cotton Exchange—were voluntary associations formed to address the public goods problems thrown up by the maturing US economy. They worked to provide space for trade; to establish rules, rates, and standards; to improve communication and the flow of information; to provide training for new workers; and to uphold professionalism among their members. Banks joined together to create nonprofit clearinghouses to provide emergency loans during financial panics. Railroad companies created industry associations that developed standards for cross-country timekeeping, for mechanical parts, and for signaling.[1] Most of the rules governing international trade are designed and enforced by the International Chamber of Commerce, a voluntary association founded in 1919. When they work, these kinds of private cooperative solutions are often faster, less costly, and more flexible than conventionally regulated alternatives.

But cooperation is fragile. Sometimes it holds, and sometimes it doesn't. In this chapter I explore the factors that make sustained cooperation possible, and those that make it fail. I suggest that even when they fail—and they often fail—cooperative efforts can lay the foundations for more robust solutions, particularly partnerships with local governments and others in pursuit of the common good. This is a story of hope followed by despair, followed by the glimmerings of renewed hope. It's difficult to be caught between a rock and a hard place, but there is sometimes a way through.

Orangutans on the Building

Gavin Neath, the chief sustainability officer at Unilever, arrived at work on Monday, April 21, 2008, looking forward to a productive day. He was surprised to discover that eight people dressed as orangutans had climbed to the twenty-three-feet-high balcony above the entrance to Unilever's London headquarters and unfurled an enormous banner proclaiming "Dove: stop destroying my rainforest."[2] The press had descended and were asking

everyone they could corner what Unilever was planning to do next. As the most senior manager on site, Neath could see a very tough day ahead.

The people in the orangutan outfits were from Greenpeace, and were protesting Unilever's use of palm oil. Unilever, Greenpeace declared, was responsible both for the destruction of the rainforest and the near extinction of the orangutans who lived in it. Cheap and versatile, palm oil is the most widely consumed oil on the planet.[3] It's in about half of all packaged products, from soap, shampoo, and lipstick to ice cream, bread, and chocolate—and it was in most of Unilever's products.[4] Demand for palm oil quintupled between 1990 and 2015 and is expected to triple again by 2050. Unilever was the world's largest buyer.

The uncontrolled production of palm oil is an environmental disaster. To clear land for palm cultivation, growers set fire to primary forests and peatlands, releasing carbon into the atmosphere at an enormous scale.[5] In 2015, Indonesia was the world's fourth-largest emitter of carbon dioxide (CO_2), after only China, the United States, and Russia.[6] The process of deforestation also pollutes local water supplies, degrades air quality, and threatens to destroy one of the world's most biologically diverse ecosystems.[7] The Sumatran orangutan has been driven to the brink of extinction.[8] In the words of one reporter: "A great tract of Earth is on fire. It looks as you might imagine hell to be. The air has turned ochre: visibility in some cities has been reduced to 30 meters. Children are being prepared for evacuation in warships; already some have choked to death. Species are going up in smoke at an untold rate. It is almost certainly the greatest environmental disaster of the 21st century—so far."[9]

The Greenpeace activists chaining themselves to the Unilever building were focused on Dove because it was one of Unilever's largest and most visible personal care brands that at the time was growing explosively. The activists were particularly furious about the fact that Unilever had played an important role in the foundation of the Roundtable for Sustainable Palm Oil four years earlier—a collection of NGOs and palm-buying companies dedicated to growing palm more sustainably—but "not a single drop" of sustainable palm oil was yet available. They accused the firm of "greenwashing" on a grand scale.[10]

Greenpeace's actions—and an accompanying series of videos that went viral on social media, gaining over two million views—forced Unilever to respond. Within a month Patrick Cescau, Unilever's CEO at the time, had publicly pledged that by 2020 Unilever would be using nothing but sustainable palm oil.[11]

The announcement got Greenpeace off Unilever's back—at least for a while—but created its own problems. No one inside the firm had a roadmap for how to pay for what might be as much as a 17 percent increase in the cost of one of Unilever's most important commodities, particularly when consumers didn't like being reminded that their lipstick (or their food) had palm oil in it in the first place.

Help arrived from an unexpected quarter. In January 2009, Cescau was replaced as CEO by Paul Polman, the first outsider in Unilever's 123-year history chosen for the top job. Paul was Dutch—a plus at a company that was jointly listed in both the United Kingdom and the Netherlands—but he had spent the first twenty-six years of his career working for Procter & Gamble, one of Unilever's largest competitors. Paul left P&G three years before to be CFO at Nestlé—another one of Unilever's key competitors—but missed out on the CEO job there in 2007. He arrived at Unilever with something to prove at a time when the firm was widely perceived as falling behind.

Unilever had been roughly the same size as P&G and Nestlé until the early 2000s, but in the five years before Paul's appointment, Nestlé and P&G had grown rapidly, while Unilever's sales had stagnated, and by 2008 Unilever's share price was less than half that of its rivals.[12] This was partly a reflection of the fact that P&G and Nestlé were active in a number of high-margin businesses (notably diapers and pet food) where Unilever didn't have a presence, and also of the fact that Unilever was weighted down by a number of notoriously low-margin products (notably margarine). But investors also believed that Unilever's organization had nothing like the focus or drive that characterized P&G and Nestlé. One observer characterized Unilever as "the basket case of the (consumer goods) industry."

The press speculated that Unilever's board had chosen Paul as the next CEO precisely because he was an outsider—and an outsider with a record of delivering bottom-line results.[13] Former colleagues suggested that he was "tough and analytic" and that he had a "tough, take-no-prisoners style."

But it turned out that Paul was a more complicated man than he first appeared. The first signal that this might be the case came on his first day as Unilever's CEO. He announced that Unilever was going to stop the practice of offering earnings guidance, telling the *Wall Street Journal*, "I discovered a long time ago that if I focus on doing the right thing for the long term to improve the lives of consumers and customers all over the world, the business results will come."[14] The share price fell 6 percent in a single day, taking nearly €2/$2.2 billion off Unilever's market capitalization. But Paul stuck to his guns, later joking that he took the plunge because "they couldn't fire me on my first day." When Neath raised the question of sustainable palm oil with him, his immediate response was "We have to do it and we can't do it alone: let's socialize the problem."

This is the central premise of industry self-regulation. If all the firms in an industry need something done—or something stopped—but are unable to address the problem by acting alone, it may be possible to solve it by agreeing to act together. In the case of palm oil, for example, every major consumer goods company in the world—many with brands worth hundreds of billions of dollars—was potentially vulnerable to NGOs accusing them of destroying the rainforest. Pepsi, for example, is one of the largest buyers of palm oil in the world.[15] So is Mars, the maker of M&Ms. Neither firm could afford a sustained campaign linking their products to pictures of orangutans being hacked to death as they ran from the flames of a burning forest.

Any single firm that chose to use sustainable palm oil not only faced the daunting challenge of finding sustainable oil to buy but also risked putting itself at a very substantial cost disadvantage. But if all the firms in an industry could be persuaded to move together, buying sustainable oil would become something that was "precompetitive," or table stakes—a cost of doing business that all firms undertook to reduce the risk of damage to their brands. If every firm in the industry agreed to buy sustainable palm oil, everyone's costs would increase—but everyone's brand would be protected, and no one would have put themselves at a competitive disadvantage.

These kinds of voluntary cooperative agreements are, of course, inherently fragile.[16] Individual firms can promise to do the right thing but fail to follow through, leaving the defectors with a short-term cost advantage, and those that chose to cooperate feeling like (angry) patsies. When I suggested to one historian of self-regulation that industry-wide cooperation might play a central role in solving the world's great problems, he giggled. His view was that industrialists have often used self-regulation merely to diffuse the threat of government regulation and to disadvantage smaller firms and potential entrants, rather than to make fundamental change, and that in general self-regulation is rarely effective except in the shadow of government regulation.[17]

But desperate times call for desperate measures. In many places, governments are corrupt and regulation is rarely enforced—and while many of our problems are global, we have few effective global regulators. Moreover there have been times and places when industry cooperation has proved to be quite successful. Consider, for example, the first attempt to clean up Chicago.

Learning from History: Black Smoke in the White City

The great industrial cities of the nineteenth century were incredibly polluted, and some of the earliest attempts at industry self-regulation were triggered by the desire to clean them up. These efforts sometimes worked—and sometimes they did not. When Chicago's business elite set out to clean up the city, for example, they initially met with considerable success.[18]

On February 24, 1890, the US Congress selected Chicago as the host city for the great "world fair" that became known as the World's Columbian Exposition.[19] New York's richest men had pledged $15 million (about $400 million in today's money) to underwrite the fair if Congress awarded it to New York City. Chicago's elite—including Marshall Field, Philip Armour, Gustavus Swift, and Cyrus McCormick—not only matched the offer but also raised several million more dollars in twenty-four hours to beat New York's bid.[20]

Hoping that the fair would bring international prominence to the city, the organizers drew up plans to build an elaborate "White City" in Jackson Park, a marshy bog seven miles outside Chicago's city limits, and hired some of the country's most prominent architects to design a suite of neoclassical buildings in the beaux arts style that would be covered with plaster of paris and painted a bright white.

As the date of the fair drew closer, many of its most ardent backers began to worry that the pristine buildings would be covered by a pall of thick, greasy smoke. Chicago—like all industrial cities of its time—was subject to appalling pollution. In the words of one historian:

Chicago world's fair at the end of the nineteenth century

> Today, one hundred years later, it is difficult to envision the sheer dirtiness and heavy blackness of the smoke that polluted the city in the early 1890s. . . . The most badly smoking buildings disgorged from their smokestacks columns of black smoke and greasy soot that reminded spectators of erupting volcanoes. Some black smoke was so heavy that it could barely float in the air. It often fell to the ground, creating almost solid banks of soot and steam and ash on city streets.

Businesspeople complained that they had to wear colored shirts and dark suits to hide the soot. Shops and factories had to keep their doors and windows tightly shut—even at the height of the summer heat—to try to prevent the smoke from damaging their goods. In 1892, J. V. Farwell, one of Chicago's leading dry goods merchants, estimated that it cost him $17,000 a year to replace goods damaged by smoke (about $430,000 in today's money). One later estimate suggested that the smoke cost Chicago more than $15 million (about $405 million in today's money) a year. And of course neither estimate includes the enormous costs that the smoke exacted in human health.

Chicago had passed the country's first anti-smoke ordinance in 1881, but it was rarely enforced. The city's Department of Health was so understaffed, there were only a handful of sanitary inspectors responsible for identifying violations, and the Attorney General's Office rarely had the capacity to bring cases. When it did bring charges, polluters would often (successfully) pressure local politicians to persuade local judges to dismiss them.

In response, in January 1892, two years after Chicago had been awarded the fair, a group of prominent Chicago businessmen formed the Society for the Prevention of Smoke, an organization dedicated to "getting rid of the smoke nuisance" before the opening of the fair, scheduled for May 1893.[21] All but one of the society's founders were directors of the exposition, and many of them were significant investors in the stocks and municipal bonds that had been used to fund it.

The group began by exhorting businesses in Chicago to clean up their smoke as a demonstration of "public spirit." Since the equipment required to prevent the generation of smoke could be difficult to install and operate effectively, the society hired five engineers—at its own expense—to publicly demonstrate the technology and to provide direct assistance to polluters. By July the engineers had sent out more than four hundred detailed reports to establishments across the city, offering specific recommendations as to how their pollution could be controlled. About 40 percent of the businesses receiving the reports implemented the recommendations and "practically cured their smoke nuisance." Another 20 percent followed the recommendations but were unable to stop polluting, and a further 40 percent refused to make the attempt.

The society next turned to the law. With the full cooperation of the city, it hired (again at its own expense) Rudolph Matz, an attorney whose job was to take owners to court if they didn't attempt to follow the recommendations they had been given by the society's engineers. Matz responded with vigor, bringing 325 suits. In just over half of the cases, the owners agreed to attempt to abate their smoke, and the charges were dismissed; while in 155 cases, the owners paid $50 fines rather than agreeing to stop polluting. In such cases Matz would often sue again. One tugboat owner, for example, estimated that he had paid more than $700 in fines because of his refusal to switch to the significantly more expensive coal required to operate without generating smoke. By late December 1892 most of Chicago's downtown smoke was under control. Roughly 300 to 325 problems had been abated, locomotive smoke had been reduced by 75 percent, and 90 to 95 percent of tugboat owners had made the necessary switch.

That spring, however, the panic of 1893 hit the city, initiating a deep depression that lasted several years. A quarter of the country's railroads went bankrupt, and in some cities unemployment among industrial workers hit 20 to 25 percent. The Society for the Prevention of Smoke began to demand jury trials, and in several high-profile cases, juries refused to convict, despite the fact that the plaintiffs were clearly making no effort to reduce their emissions. We can't tell from the public record exactly why this is the case, but one possibility is that the members of the society were increasingly seen as "fat cats" attempting to control city government for their own ends. Believing that without public support their cause was hopeless, the society formally disbanded in 1893.

In the event, the fair proved to be an enormous success, setting a world record for outdoor event attendance and seeing more than

750,000 visitors. The contrast between the gleaming white structures of the fair and the filth of downtown Chicago has been credited with helping to start the civic improvement movement that emerged in the last decade of the century—a movement inspired by the idea that cities could be as clean and healthy as the White City. But Chicago didn't effectively address its air pollution problems until the 1960s.

Building Cooperation at Global Scale

So far, so good. The White City case is encouraging. But it describes efforts within a relatively small, tightly knit community with very compelling reasons to cooperate. Can cooperation be sustained in a much more global, much more diffuse setting—like palm oil? The answer to this turns out to be complicated. Five years ago I—and many people within the industry—thought of the palm oil case as one of the great examples of successful cooperation in the service of the common good. Today it's clear that this verdict was premature.

Unilever succeeded in socializing its problem—the vast majority of its competitors have agreed to switch to sustainable oil, and Unilever is on target to use 100 percent sustainable oil by 2019, a year ahead of its commitment. But palm oil cultivation continues to be a major driver of deforestation. It has become clear that the only way to solve this problem is in partnership—with investors, with local communities, and with local governments. The industry's self-regulatory efforts have increased the odds that these partnerships will succeed, but the situation is still very much in flux.

I begin this section by describing how these dynamics have unfolded, since they are a powerful source of insight into the opportunities and the threats that constrain many of the global efforts that are currently underway. I then turn to a discussion of the success of the beef and soy initiatives, suggesting that their ability to partner with local regulators was fundamental to their success. There are hundreds of global self-regulatory efforts currently underway, attempting to solve problems as diverse as ocean pollution, overfishing, corruption, and abusive labor conditions across almost every industry. Building a richer understanding of the likely determinants of their success is crucial if we are to reimagine capitalism.

Paul began his efforts to socialize his palm oil problem by reaching out to members of the Consumer Goods Forum (CGF), one of the largest industry associations in the world. The forum currently includes more than four hundred consumer goods manufacturers and retailers from seventy countries. Between them, they generate more than $3.87/€3.50 trillion in revenues and employ nearly ten million people.[22]

In early 2010, in a series of small group meetings with fellow CEOs, Paul began to advocate for the idea that stopping deforestation should be a key issue for the forum. His efforts were greatly helped by a Greenpeace attack on Nestlé. In March 2010 Greenpeace released a spoof ad, showing a bored office worker biting into a KitKat and finding himself eating an orangutan's bloody finger instead. (You can see it on YouTube, but be warned—it's not pretty.[23]) The intense press attention that followed galvanized not only Nestlé but also many of the other consumer goods companies. Scott Poynton, the director of one of the NGOs that Nestlé had hired to grapple with the issue, remembers arriving at Nestlé's corporate headquarters to be greeted by the receptionist saying plaintively, "We don't want to kill orangutans, that's not who we are."[24]

Gavin and Paul introduced the group to Jason Clay of the World Wildlife Fund, who argued that the road to sustainability lay in precompetitive cooperation by a small number of major companies. He pointed out that in all of the world's most highly traded commodities, a hundred companies bought at least 25 percent of the world's production. He suggested if these companies demanded that the commodities they bought be sustainably grown, entire industries would be forced to move in a more sustainable direction—and that persuading a hundred companies to act would be much easier than persuading 25 percent of the world's consumers to do so.

Gavin remembers one meeting in particular—a small gathering at Unilever's headquarters that included the CEOs of fifteen of the world's largest consumer goods companies, including Nestlé, Tesco, P&G, Walmart, Coke, and Pepsi—as "a magic moment." Terry Leahy, the CEO of Tesco, at the time the third-largest retailer in the world, suggested focusing on sustainability "through a carbon lens"—and was met with enthusiastic approval. Several of the CEOs present took on persuading their peers to address deforestation as a personal mission.

In the months that followed the group struggled to get the other members of the forum on board. I'm told it was a ferociously difficult process, and antitrust concerns meant that the minutes of every meeting and all the documents they generated had to be scrutinized by antitrust lawyers. Nevertheless, the steering group charged with putting together a concrete proposal around Leahy's idea ultimately reached an agreement. At an emotionally charged meeting of the forum, Paul, together with the CEOs of Tesco, Coca-Cola, and Walmart, gave their full-throated support to the proposal, arguing passionately for the other CEOs in the room to join them. In November 2010, at the UN's sixteenth climate conference, Muhtar Kent, Coca-Cola's CEO, announced that the members of the forum were committing to achieving zero net deforestation by 2020 for the four commodities most responsible for driving global deforestation: soy, paper and board, beef, and palm oil.[25] Paul and his colleagues had succeeded in persuading nearly every major Western consumer goods company and nearly every major retailer to commit to buying and selling only sustainable palm oil—defined as deforestation-free palm oil grown under well-regulated labor conditions.

But this was just the first step. Self-regulation is only stable when all the parties to an agreement believe that it is in their collective interest to cooperate. But while this is a necessary condition it is not sufficient. For cooperation to endure it must also be the case that the participants cannot easily "free ride," by example, promising to use sustainable oil but not actually doing so—and for this to be the case the group must have the ability to know when a firm cheats and the ability to sanction or punish such firms if they are caught.

Rather than focus attention on the consumer goods firms themselves, Paul and his colleagues began by trying to create a reliably sustainable supply, reasoning that it would then be relatively easy to observe whether firms were buying from it. As a first step they began by focusing on the three firms that handled the vast majority of the international *trade* in palm oil: Golden Agri-Resources (GAR), which in addition to its trading business was also the largest grower of palm in Indonesia; Wilmar, the approximately $30 billion agricultural giant based in Singapore that handled almost half of all globally traded palm oil; and Cargill, a privately held American company that was the world's largest trader in agricultural commodities, with over $100 billion in revenues. They believed that if they could persuade the three firms to commit to zero deforestation, they would together push a large majority of palm oil suppliers toward sustainability—and the regular use of sustainable certification.

GAR had adopted a zero burning policy in 1997, but had continued to carry out forest clearance without permits, burning and disturbing areas of deep peat and thus releasing enormous amounts of carbon. At the end of 2009, despite significant concerns over the commercial impact of the decision, Unilever announced that it would no longer buy from GAR unless the firm changed its practices.[26] The move sent shockwaves through the palm oil industry, triggering riots and demonstrations in Indonesia. But in 2010, Nestlé joined Unilever in pressuring the firm, and Kraft and P&G quickly followed suit. GAR reached out to Greenpeace, opening negotiations that continued over a tense year. (One observer described the atmosphere as "worse than that between the Arabs and the Israelis.") In February 2011, GAR pledged not to clear high conservation value (HCV) forests and peatlands, and to refrain from clearing forest areas storing large amounts of carbon. The four companies then resumed doing business with the firm. When asked why GAR became the first Indonesian palm oil company to announce a Forest Conservation Policy, Agus Purnomo, the chief sustainability officer of GAR, said,

> Because our primary market, the premium buyers, were requesting us to do it. Is it because we want to go to heaven? No. Of course everybody wants to go to heaven, but we are doing it because our buyers asked us to do it. It is what every company needs to do, to (fully satisfy) their customer.

At the same time, members of the CGF began to reach out to Wilmar and Cargill in an attempt to persuade them to change their sourcing policies, complementing the efforts of a number of NGOs that had been targeting Wilmar for years. Fortuitously, in June 2013 thick layers of soot and haze caused by illegal wildfires burning in Indonesia blanketed Singapore, Wilmar's hometown. The miasma set records for air pollution in the city-state and blanketed cars with ash, forcing people indoors. The press attention led Wilmar's CEO, Kuok Khoon Hong, to engage directly with Paul and with both Forest Heroes and the Forest Trust, two of the leading NGOs in the palm oil space. "He talked a lot about how upset he was at the haze in Singapore and in China," one of the activists recalled. "He just needed to be given a business rationale to go ahead." In December 2013, Wilmar signed a sweeping "No Deforestation, No Peat, No Exploitation" pact,[27] and in July 2014 Cargill, the third major palm oil trader, released an updated policy committing to deforestation-free, socially responsible palm oil.[28]

Translating these commitments into action on the ground was the next hurdle. The first point of contention was to define exactly what counted as "sustainable" palm oil. It was relatively easy, for example, to agree that palm oil grown on land that had high conservation value forests on it the year before was not sustainable. But what was a high conservation value forest, and who was to define it? Did it count as deforestation if the land was a second-growth forest? What kinds of labor conditions made a particular plantation sustainable?

One option was to use the standards developed by the Roundtable on Sustainable Palm Oil (RSPO), a multistakeholder partnership that had been founded in 2004 to develop standards for sustainable oil palm cultivation. "In the beginning it was very, very tough," Roundtable CEO Darrell Webber explained:

> The seven stakeholder groups from the supply chain, which included several environmental and social NGOs, came together and basically no one trusted each other. There was lots of heated debate, lots of arguments. It took more than a year to draft the first standard. Many close calls, threats of walkouts and dissatisfaction. But at the end of the day, trust was built. People started to understand the other parties' views much better over time.

RSPO issued the first global guidelines for producing sustainable palm oil in 2005. The guidelines were articulated in eight principles and forty-three "practical criteria" that were intended to be revised every five years, and were adapted for use in each country. The growers, who paid for the audits, were assessed for certification every five years and, if certified, monitored annually.[29] All the organizations that took ownership of RSPO-certified oil palm products were required to be supply chain certified and could then use the RSPO trademark. The certification remained entirely voluntary, but it could be withdrawn at any time in the case of infringement.

But critics claimed that the RSPO's standards were comparatively weak and that it was slow to respond to new developments. In 2015, a comprehensive report summarized a list of breakdowns in oversight, including fraudulent assessments that covered up violations of RSPO standards, failures to identify indigenous land right claims, labor abuses, and conflicts of interest due to links between the certification bodies and plantation companies.[30] In response, a number of individual growers—often under pressure from Western buyers—agreed to use more stringent standards. Western buyers also continued to put pressure on the RSPO to tighten its criteria. Industry participants described this process as one of attempting to continually move the "floor"—or the minimum requirements, as captured by RSPO standards—while also continually pushing the "ceiling" or the definition of sustainability, based on the best available knowledge, at the same time.

The foundations had been laid for third-party certification of sustainable palm oil, making it relatively easy to tell if the big consumer goods companies were keeping their commitments, and the technology in this area continued to improve. There were important advances in the ability to audit any single firm's supply chain—to track oil back to the mill where it was processed and to monitor the plantations where it was grown. Wilmar, for example, was regularly deploying drones to help ensure that its plantations were indeed being sustainably managed. Given the strength of the economic case for switching and the readiness of many NGOs to call out firms that didn't do the right thing, many people—including both me and Jeff Seabright, who had taken over from Gavin Neath as Unilever's chief sustainability officer—were confident that the Consumer Goods Forum's actions would dramatically slow palm oil–related deforestation.

But between 2001 and 2012 deforestation rates in Indonesia, the largest producer of palm oil, more than doubled,[31] falling only slightly between 2012 and 2015 and increasing significantly in 2016. Rates fell again in 2018, but Indonesia is still losing hundreds of square miles of forest cover every year.[32] Between 2010 and 2018, the country lost nearly five thousand square miles, equivalent to 480 Mt of CO_2 emissions, and 27 percent of this came from primary rainforests.[33] The share of the world's palm oil that is sustainably grown has not budged since 2015, and it is now clear that many members of the Consumer Goods Forum are not going to be able to meet their 2020 commitments.[34] Industry-wide cooperation has enabled many of the Western firms to meet their promise to use sustainable oil—but it has not fixed the underlying problem.

A number of factors appear to be responsible for this outcome. The first is the failure to anticipate that while the business case for switching to sustainable oil was relatively strong for the big Western buyers and the large palm oil suppliers, persuading the small farmers who grow nearly 40 percent of the palm oil crop—and who were responsible for much of the deforestation and the burning that occurred—to stop cutting down the forest was another challenge again.[35] Two hectares (one hectare equals 2.47 acres) of cleared rainforest planted in palm could ensure a family's future, providing enough income to send the children to university. Moreover, programs designed to support smallholder farmers become more sustainable were having only mixed success.

Smallholders typically achieve yields of less than two metric tons per hectare, compared to the six to seven metric tons reached by best-practice plantations, so increasing smallholder efficiency is one possible solution to this problem. But improving smallholder productivity is difficult. It requires educating hundreds of thousands of smallholders in new planting and harvesting practices, in regions that lack the kind of intermediate cooperative structures that were so helpful in transforming tea production. The smallholders also had to be financed—first with higher-quality seed and equipment, and then throughout the first, nonproductive years of oil palm growth. While Cargill believed that improving smallholder productivity was "the only way forward" and claimed that their preliminary efforts were proving to be quite successful, GAR, which had piloted a project to fund and educate smallholders in Kalimantan, part of the Indonesian segment of the island of Borneo, was less optimistic. During the pilot, some smallholders had failed to farm sustainably, while others had sold their harvest to independent mills that promised a higher price than GAR.[36]

The legal and political environment was another major challenge. Over 90 percent of the world's palm oil is grown in Indonesia and Malaysia, and in both countries it is an important pillar of the economy. In 2014, for example, agriculture made up over 13 percent of Indonesian gross domestic product[37] and employed over 34 percent of the population,[38] providing two-thirds of rural household income to approximately three million people. Palm oil was Indonesia's second-largest agricultural product and the country's most valuable agricultural export.[39] In Malaysia, agriculture accounted for 7.7 percent of the nation's GDP.[40] Many politicians in both Indonesia and Malaysia believed there to be a direct conflict between local economic development and sustainability.

Worse, Indonesian law required land concession holders to develop *all* of the land they had been allocated, regardless of company policy, and different Indonesian ministries used different maps. Kuntoro Mangkusubroto, the minister responsible for the effort explained,

> Every important ministry has their own map of Indonesia. Well, Indonesia is a very big country, and these ministries each have their own mission, so it may make sense that they have their own

> version. But when it comes to national development, we need one map. We need the map to have a conclusion that can be accepted by the public, by the politicians, by the government, that, for example, the forest coverage is so much, how many million hectares on this island, where the boundary of the forests is.

The use of a single official map could help reduce the issuance of overlapping permits, a leading cause of disputes between companies and indigenous people. In 2010, the Indonesian government announced a One Map initiative, aiming to bring together spatial data for Indonesia into a single database, but the project—which is partially funded by the World Bank—is still in process.[41]

Another major problem is that there is simply too much money to be made in cutting down Indonesia's forests. An increasing fraction of globally traded palm oil is sold to Indian and Chinese firms, and few of them have shown any interest in purchasing sustainable oil. Moreover, although some Indonesian administrations have committed themselves to the goal of reducing deforestation, the palm oil industry is an important source of patronage for both local and national politicians.[42] The Ministry of Forests, which has partial responsibility for land use and allocation, is notoriously corrupt, and retiring civil servants often buy a palm oil mill or two "for their retirement." Patronage networks are endemic to Indonesia and often rely on the revenues from deforestation and illegal logging to grease the wheels of politics. It was difficult to know how the CGF could address these kinds of issues. One experienced NGO leader put it to me this way: "Suppose that you detect illegal logging from the air. You call the local mill, and then ask them to do what? To drive into the forest, confront the armed men who are guarding the site, and tell them that the mill will refuse to purchase palm oil produced from the land six years from now?! I'll tell you what you do when you encounter illegal loggers. You smile, wish them well and keep on your way."

More than 70 percent of all the logging in Indonesia is illegal.[43] If there's an important group of producers who see no benefit to becoming more sustainable, if they have customers who are happy to buy from them, and if the government is not willing to enforce its own laws, it's going to be very difficult to stop the forests coming down.

Does this mean that self-regulation has failed in palm oil? The effort has—so far—failed to stop deforestation. But it has significantly increased the odds of stopping it going forward—if a way can be found to bring governments and/or investors into the mix. The coalition needs to find a way to increase everyone's incentives to cooperate—by, for example, making growing sustainable oil economically attractive to smallholders, by persuading Indian and Chinese consumers to push local firms to use sustainable oil, or by persuading local governments to enforce laws against deforestation.

The business case for action remains strong, and years of work have generated deep knowledge as to how to fix problems on the ground. But someone needs to be able to enforce cooperation. The decade-long struggle to reduce the deforestation associated with soy and beef production in the Amazon suggests that the secret is to partner with the public sector.

The soy story begins on familiar ground. In 2006 Greenpeace published "Eating up the Amazon," a report claiming that Archer Daniels Midland (ADM), Bunge, and Cargill—the world's largest commodities trading companies—were actively contributing to the destruction of the Amazon rainforest through their financing of soy bean production.[44] Deploying protestors dressed in seven-foot-high chicken suits outside McDonald's (95 percent of soy is used as animal feed), Greenpeace accused the Western firms buying Brazilian soy of helping to destroy one of the world's last great rainforests and cooking the planet.

Greenpeace published its report (and let loose its chickens) on April 6, demanding that the entire food industry exclude soy

produced in the Amazon from their supply chains. Three months later—on July 25—a group that included not only ADM, Bunge, and Cargill but also McDonalds and the two Brazilian industry associations that controlled 92 percent of Brazilian soy production announced a "soy moratorium"—an agreement not to purchase soy grown on lands deforested after July 2006 in the Brazilian Amazon.

The moratorium was monitored by the Soya Working Group, which included soy traders, producers, NGOs, purchasers, and the Brazilian government. Using a combination satellite/airborne monitoring system developed jointly by industry, several NGOS, and the government, the Soya Working Group monitored 76 municipalities that between them were responsible for 98 percent of soy production in the Amazon.[45] Farmers who violated the moratorium were prevented from selling to the moratorium's signatories, and could find it difficult to obtain financing. The agreement was renewed every two years, and, in 2016, the moratorium was extended indefinitely, or until it was no longer needed.[46] In the ten years after it was signed, soy production in the Brazilian Amazon nearly doubled,[47] but less than 1 percent of the new production was on newly deforested land.[48] Soy yields—the quantity of soy grown per acre—also increased significantly.[49]

In 2009 Greenpeace issued "Slaughtering the Amazon," a report accusing the cattle industry of clear-cutting mature forests in the Amazon.[50] Nearly 60 percent of agricultural land worldwide is used for beef production, and cattle production drives 80 percent of Amazonian deforestation.[51] Federal prosecutors in the Brazilian state of Para began suing ranchers who had illegally cleared forestlands and threatening to sue retailers who were buying from them. In response Adidas, Nike, Timberland, and a number of other shoe companies using Brazilian leather announced that they would cancel their contracts unless they could be assured that the leather they were using was not implicated in the destruction of the Amazon. The Brazilian Association of Supermarkets called for the beef they sold to be deforestation-free.

The shares of Brazil's four-largest meatpackers fell significantly as a result.[52] Together they signed what became known as "The Cattle Agreement," banning the purchase of cattle from newly deforested areas in the Amazon.[53] Continued customer pressure—including a 2010 commitment by the members of the Consumer Goods Forum to buy only zero deforestation beef—helped to keep the moratorium in place.

Here again it was the active support of the Brazilian government that was particularly helpful. Most of the Brazilian Amazon is formally protected by the Forest Code, which requires that landowners permanently maintain 80 percent of their land as forest. It was passed in 1965 but only rarely enforced until the 2010s, when the combination of the soy and beef moratoria and the development of sophisticated technology for tracking deforestation on a daily basis gave it new life. The Cattle Agreement was remarkably successful. In 2013, 96 percent of all slaughterhouse transactions were with suppliers registered with Brazil's Rural Environmental Registry, up from just 2 percent of transactions prior to the agreement.[54] Between them, the two agreements dramatically slowed deforestation in the Amazon at a time when rates of deforestation increased significantly nearly everywhere else in the world.[55]

In both cases government support was critical to making progress—as the dramatically accelerating rate of deforestation in the Amazon following President Jair Bolsonaro's recent election and subsequent repudiation of his predecessor's policies makes only too clear.[56] But in both cases government support was catalyzed and enabled by private sector action. The industry's commitment gave the government political cover to enforce the law—and provided critical technical know-how and ongoing support.

My guess is that this experience will prove to be the model for future self-regulatory efforts. Industry-wide cooperation will establish a demand for sustainably produced products. Leading firms will invest to build the technical expertise and operational sophistication necessary to make the switch. But in the end government support will be critical to making progress.

One study, for example, analyzed the effectiveness of private sector regulation in the apparel and electronics industries. It drew on five years of research, more than 700 interviews, visits to 120 factories, and extensive quantitative data.[57] The author concluded that while there was much that could be done, private sector compliance programs were unlikely to solve the full set of labor problems in the global supply chain. In his words:

> After more than a decade of concerted efforts by global brands and labor rights NGOs alike, private compliance programs appear largely unable to deliver on their promise of sustained improvements in labor standards. . . . Compliance efforts have delivered some improvements in working conditions . . . (but) these improvements seem to have hit a ceiling: basic improvements have been achieved in some areas (e.g.: health and safety) but not in

> others (e.g. freedom of association, excessive working hours). Moreover, these improvements appear to be unstable in the sense that many factories cycle in and out of compliance over time.

In palm oil and in textiles, extensive and well-funded efforts at self-regulation have succeeded in making major gains but have not achieved their original objectives. In both cases the industry has begun to look to local regulatory authorities as partners in achieving fully sustainable supply chains.

In palm oil, members of the Consumer Goods Forum have been meeting regularly with a broad range of stakeholders—including NGOs and local communities and politicians in Indonesia and Malaysia—to explore possible ways forward. One possibility is to move to what's technically known as a "jurisdictional" approach—building partnerships with local politicians, local NGOs, and local communities in an attempt to build a business case for converting entire regions to sustainable palm. Similar conversations are happening in the textile business in the context of some promising early success. One study of the Indonesia apparel industry, for example, found that self-regulatory efforts were significantly more likely to increase wages when the self-regulating body worked closely with the state and when local unions were mobilized to push for state action.[58] A study of the Brazilian sugar industry found that the efforts of private auditors complemented the attempts of local regulators to prohibit extreme forms of outsourcing, and that the two together pushed firms to adopt significantly improved labor standards.[59]

What Makes the Difference?

What makes the difference? Why do some self-regulating organizations succeed while others fail? One answer to this question emerges from the history of the Institute of Nuclear Power

Operations.[60] The institute was founded in 1979 following the disastrous nuclear reactor meltdown known as Three Mile Island. The accident shocked the public and terrified the nuclear industry. Indeed many of the utilities operating nuclear power plants became convinced that the industry could not survive another such incident.[61]

Historically the nuclear power industry had been regulated by the Nuclear Regulatory Commission, a US government agency. Following the Three Mile Island disaster, the Nuclear Regulatory Commission attempted to increase the industry's safety, but it was essentially a technologically focused institution, and the independent commission set up to investigate the accident concluded that organizational and managerial issues such as complacency and miscommunication were the primary cause of the accident, rather than problems with the technology. Many nuclear power workers had gained their experience working with fossil fuels and had assumed that they should run nuclear plants as they had run fossil fuel plants, namely, as hard as possible until they hit problems—problems that would then be fixed by maintenance and/or technical crews. Many managers and operators seemed to lack a sense of the vastly greater destructive potential of nuclear energy. When individual plants did learn something about how to run a plant more safely, the information was not shared with other firms. The fifty-five utilities operating nuclear power plants in the United States set up the Institute of Nuclear Power Operations as a private self-regulatory organization to fill this gap.

The institute was staffed by former nuclear Navy veterans. The Navy's nuclear program was famous for its zero accident record and a culture that placed safety first, second, and third. The Navy men (they were all men) developed operating standards and procedures for the industry and provided operational support for their adoption through an aggressive program of training and plant visits. Each plant was extensively evaluated each year. Following the visit, institute employees would show each plant how it compared to its peers across a set of critical performance indicators—and then offer to work with the plant to bring performance up to par. At annual industry meetings, the institute would present the results of these studies to all attending CEOs, putting further pressure on those with low grades to address their plants' problems. If an executive was found to be uncooperative, the institute could also threaten to contact the company's board of directors.

Between 1980 and 1990, the average rate of emergency plant shutdowns fell more than fourfold, and the institute is widely credited with making an order of magnitude improvement in the safety of the US nuclear power industry. It is still in operation today and continues to be entirely funded by the nuclear industry.

As I mentioned above, Elinor Ostrom's pioneering work also uncovered many examples of successful industry-wide cooperation. In one of her most famous studies, she examined the Maine lobster industry. Lobster stocks in Maine declined dramatically in the 1920s and 1930s; in response the state imposed regulations on the size and number of lobsters that could be taken. Local lobstermen then self-organized to enforce these limits. They agreed to throw back breeding females after punching a notch in their tails, and established a system for dividing the fishing ground among themselves and an enforcement mechanism to prevent violations. Lobster stocks were back to sustainable levels by the late twentieth century and are now booming.[62]

Together with the Chicago story, these two cases graphically illustrate the four conditions that must be in place if self-regulation is to succeed. The first is that sustaining cooperation must be in everybody's interest—and must clearly be seen to be so by everyone involved. It's much easier to cooperate if doing so yields immediate benefits and—equally—if the costs of failing to cooperate are also significant. One of the reasons the nuclear utilities were so eager to cooperate after the Three Mile Island disaster was that they feared that a single slipup at *any* nuclear plant could put the entire industry out of business.[63] They thus had very strong incentives to make cooperation work. This was also the case in the lobster industry, where continuing to overfish would certainly throw everyone out of work, while reducing the catch would plausibly lead to a rapid rebound in the fisheries. One of the reasons that nearly half of the world's fisheries are sustainably fished is because most fisheries have historically rebounded relatively quickly once fishing has been controlled. It means that fishers don't have to wait long to see the benefits of their self-restraint.[64] In the Chicago case, on the other hand, it's not surprising that the Society for the Prevention of Smoke was almost entirely founded by men who had significant financial stakes in the success of the White City, while it was the tugboat owners—who arguably had the most to lose and the least to gain from changing their behavior—who made achieving lasting cooperation so difficult.

Cooperation is also much easier if everyone involved in the industry is in it for the long term, or, more technically, if it's hard to enter and difficult to leave. Again, this was clearly the case in both the nuclear and lobster cases. Nuclear plants have a sixty-year life and cannot be moved. The lobster fishers had gone into debt to buy boats and equipment—assets whose value would fall close to zero if the fishery collapsed.

But these two conditions are only enough to ensure everyone will cooperate if the benefits of doing so ensure that it's in no one's interest to cheat or free ride. One of the reasons that voluntary bodies like the International Chamber of Commerce are so often successful is that the benefits they offer are tangible and immediate and the temptation to cheat is very small. When this isn't the case, cooperation will only survive if it's easy to see if someone is not pulling their weight. In the nuclear case, annual inspections by the Institute of Nuclear Power Operations served not only to bring every plant up to speed with the latest techniques but also to ensure that all the utilities were doing their best to use them. Lobster fishing catches are harder to observe, but the small size of the lobstering communities made it relatively easy to detect people who were cheating.

The fourth and last condition is that it must be relatively easy to punish those members of the coalition who are not playing by the rules. The nuclear guys became very good at this. In one famous incident, the institute sent a letter to the board of directors of Philadelphia Electric's Peach Bottom plant, highlighting the plant's years of underperformance. The board retired the plant's top executives, including the CEO, and moved quickly to address the problems. In another incident, after years working privately with the management to fix California's Rancho Seco nuclear reactor, the institute informed the government's nuclear regulators about the reactor's numerous safety violations. The regulators then conducted their own inspection of the plant and subsequently ordered it shut down.

In the lobstering case, poachers who left their lobster traps in another lobsterman's territory could expect a series of gradually escalating sanctions. A tag would be tied to the poacher's trap to signal that he had been caught. If the poaching persisted, other lobstermen might cut the rope leading from the buoy to the trap, making the trap impossible to retrieve. Poachers who persisted could expect damage to their boats or threatening visits to their homes.

Notice that it was the loss of the ability to punish persistent polluters who continued to pollute that ultimately destroyed the Chicago coalition. As long as public opinion was with the coalition and the courts were convicting polluters, the vast majority of Chicago's businesses would stay in line. But once opinion turned against the coalition and no one would convict, the effort fell apart.

Cooperation Within Regions

The idea that collective action across firms might be a powerful first step toward partnership with local governments is not, of course, a new idea. Leading-edge companies have been working together with local regulators and local communities to create public goods that benefit the entire community for at least a hundred years.

In Minneapolis–St. Paul, for example, which is by some measures one of the most successful cities in the United States, the business community has a long history of working with local government, particularly with regards to education. Despite the fact that the city is located thousands of miles from either coast and has some of the worst weather in the United States, nineteen of the Fortune 500—including United Health Group, 3M, Target, Best Buy, and General Mills—are headquartered in the city, as is Cargill, the largest privately owned company in the United States. Given the city's geographic isolation and terrible weather, the CEOs of these companies are very much aware that they have a common interest in making the region an attractive place to live and work. They also have a long history of developing the kind of shared identity and private meeting spaces that help sustain cooperation.[65]

For example, according to Robin Johnson, the president of the Cargill Foundation, "The physical climate and location,[66] the isolation of the community from the coasts and the work ethic of the Scandinavian and German immigrants who settled here may have prompted the idea that we've got to do things to build the community for ourselves. No one will do it for us. We've got to do things together."

Kendall Powell, ex-CEO of General Mills, expanded:

> If you go back far enough, the Cargills and Macmillans of Cargill, George Pillsbury at Pillsbury, Cadwallader Washburn at General Mills, the Daytons of Dayton-Hudson (later Target), they lived here and managed these companies well into their maturity. As a result, organizations like the Minneapolis club became places where it was relatively easy, if there was a community problem,

> to convene a half a dozen business leaders and decide what the business community ought to do about it. The heads of these organizations now come from all over, but the institutions and the traditions that gave rise to that sense of community involvement are still there, and are still consciously pursued by community leaders.[67]

The Minnesota Early Learning Foundation is one example of this cooperation in action. It got its start in 2003, when Art Rolnick, the head of research at the Federal Reserve Bank of Minneapolis, published a paper noting that fewer than half of Minnesota's incoming kindergartners were emotionally, cognitively, or socially ready for school.[68] Cargill's CEO, George Staley, took the lead in asking the local business community for funds to take action on the problem. By the end of the 2008 recession, he had raised $24 million and persuaded the CEOs of Ecolab, Target, and General Mills to commit to five years of quarterly meetings—in person—as board members.

They used the money to test three complementary initiatives. They gave qualifying parents annual scholarships of up to $13,000 to spend on any high-quality early childcare education programs in the Minneapolis–St. Paul area. They launched a "Parent Aware" ranking system to identify high-quality early childhood education programs, and they supported every family in the program through home visits. The effort proved to be very successful, with the scholarship recipients showing significantly better outcomes than a comparison group, and it led to both the state and the federal governments making major commitments to early childhood education in the state.

Looking back, Charlie Weaver, the CEO of Ecolab, stressed the way in which the private sector was able to support investment in innovation and experimentation that could subsequently provide a basis for political action, saying:

> The biggest success was in raising the $24 million. Without the funds to pay for the scholarship, set up the rating system and give the parents the chance to choose quality daycare facilities, the idea would not have been taste tested and quality daycare centers would not have opened in these neighborhoods. Without the money we would just have done a report saying that early education is important and had no further impact. The most important thing was being able to prove the idea before taking it to the legislature for broader support.

Several mayors have told me that Minneapolis–St. Paul has a number of things going for it that have made it particularly easy to build this kind of cooperation. It is, for example, ethnically and racially much more homogenous than most American cities. But my sense is that there are literally hundreds—perhaps thousands—of city and regionally based efforts underway featuring some form of public-private partnership in pursuit of reducing environmental damage and/or reducing inequality through increased economic growth. All these efforts require at least some level of cooperation among the leading businesses of the area to be effective. Self-regulation—particularly when it's coupled with an appreciation for the power and role of the state—may yet turn out to be a crucial tool in reimagining capitalism.

Investors as Enforcers

Cooperation among investors is another key to progress. More than a third of the world's invested capital—about $19 trillion—is controlled by the world's hundred largest asset owners. Nearly two-thirds of this money is in pension funds, while the remaining third is in sovereign wealth funds.[69] The fifteen largest asset managers collectively handle nearly half of the world's invested capital. They include BlackRock, which currently manages just under $7.0 trillion;

the Vanguard Group, which controls $4.5 trillion; and State Street, which has $2.5 trillion under management.[70] A very high proportion of this money, as we saw earlier in the chapter on rewiring finance, is in passive investments. In the United States, for example, 65–70 percent of all equities are held by index and quasi-index funds.[71] These investments are completely exposed to system-wide risk. Their owners cannot diversify away from the risks that accelerating rates of environmental degradation and inequality present to the entire economy. The best way to improve their performance is to improve the performance of the economy as a whole.

In principle these investors have enormous power to move the entire economy in more sustainable directions. All they have to do is find a way to cooperate. If all fifteen asset managers or all one hundred asset owners decided together to require all the companies in their portfolio—or all the companies in a particular industry—to move away from fossil fuels, end deforestation, or embrace high road labor strategies, it would be an enormous step toward building more equitable and sustainable societies. Hiro was able to drive significant change in Japan on the basis of owning 7 percent of Japanese equities. Imagine what might happen if the owners of more than half of the world's capital demanded change. It's not quite that easy, of course. Take, for example, the ongoing effort to use investor power to arrest global warming.

Climate Action 100+ (CA100+) was founded in 2017 with the goal of persuading the world's one hundred most important carbon emitters to, as one reporter put it, "cut the financial risk associated with catastrophe."[72] The group is an affiliation of more than three hundred investors who between them control nearly half the world's invested capital.[73] They have three goals. The first is ensuring that every firm in which they invest has a board-level process in place to evaluate the firm's climate risk and to oversee plans for dealing with it. The second is to have every company clearly disclose these risks, while the third is to persuade each firm to take action to reduce GHG emissions across its value chain rapidly enough to be consistent with the Paris Agreement's goal of limiting global average temperature increase to well below 2°C.[74]

The business case for participating in CA100+ is well defined: the investors who have joined it believe that climate change presents a clear and present danger to the long-term value of their investments from which they cannot diversify away. Like Hiro, many believe that their fiduciary duty to their beneficiaries requires them to do everything they can to solve global warming. This doesn't mean that coordinating the group is entirely easy.

Its actual work is done through a mix of public letters, formal and informal conversations with company management, and the filing of "shareholder resolutions"—investor proposals for action that are submitted to a vote of the entire shareholder base at the company's annual general meeting. Individual investors take responsibility for coordinating action with respect to a particular company, building a coalition among the company's investors to press for change.

For example, in December 2018, a group of investors representing more than $11 trillion published a letter in the *Financial Times*, saying, in part:

> We require power companies, including power generators, grid operators and distributors, to plan for their future in a net-zero carbon economy. Specifically, we request companies to set out transition plans consistent with the goal of the Paris Agreement, including compatibility of capital expenditure plans. We expect explicit timelines and commitments for the rapid elimination of coal use by utilities in EU and OECD countries by no later than 2030, defining how companies will manage near-future write downs from fossil fuel infrastructure.[75]

Six months later investors from CA100+ pushed Shell into announcing short-term targets for limiting greenhouse gas emissions, and persuaded BP to support a shareholder resolution that binds the company to disclose the carbon intensity of its products, the methodology it uses to consider the climate impact of new investments, and the company's plans for setting and measuring emissions targets. The resolution also requires annual reporting on the company's progress toward these goals and an explanation of the degree to which executive pay is linked to the firm's ability to meet them.[76] The filing of shareholder resolutions might not sound like world-changing stuff, but such resolutions can be a powerful way of communicating investor priorities and beliefs and of putting pressure on the firm. Any management team is only too aware that a sufficiently large coalition of investors can, after all, replace them.

But engaging with companies in this way is costly, and Climate Action 100+ faces a classic free-riding problem: there's a real risk that any particular investor will be tempted to let the other investors in the coalition do all the hard work. It's relatively easy to tell which investors are pulling their weight, but there's no foolproof way of punishing those who choose to stay on the sidelines despite their commitments. My sense is that at the moment, leading members are using a mixture of moral suasion and in-group shaming to persuade everyone to participate. If they succeed, Climate Action 100+ will prove to have been the tip of a critically important iceberg.

I read BlackRock CEO Larry Fink's letter and the announcement by the Business Roundtable of a new purpose for the corporation as testing their colleagues' appetite for exactly this kind of strategy, and a group of the world's hundred largest investors—or of the world's fifteen largest asset managers—would meet several of the conditions most likely to support cooperation. The group would be relatively small, the returns to cooperation potentially very high, and it should be straightforwardly easy to observe whether every member of the group is indeed putting pressure on the firms they own. If the coalition could develop a way to punish investors that free ride, they would be home free—and it might be that social pressure within the group would be enough to persuade everyone to cooperate. I'm told that there is a room in which many of them meet. The rumor is that they look at each other and say, "You go first."

My students ask me if this kind of thing doesn't make me nervous. Do we really want the world's largest asset owners to exercise this kind of collective power? To me, this is an easy question. These owners are already exercising enormous power—in the service of pushing the firms in their portfolio to race to the bottom. It's critically important that they make a deliberate decision to trigger a race to the top instead. A central element of a reimagined capitalism is a reimagined financial sector—one that takes its collective responsibilities to the world seriously and is willing to act on them.

IN SUMMARY, SELF-REGULATION is a potentially powerful way to mobilize the world's business community in support of creating collective shared value. In the nuclear industry, in textiles, palm oil, beef, and soy, and in places like Minneapolis–St. Paul, the business community has come to believe that it would be better off if it acted together to create a public good (or cease creating a public bad). This kind of cooperation keeps bumping into government, and one of the reasons that the pursuit of industry-level cooperation is potentially so important is that it creates an appetite for government intervention. In palm oil, soy, and beef, and increasingly in the textile and IT businesses, industry leaders are actively pushing for government regulation. Firms that have committed themselves to behaving well have strong incentives to push for sanctions against those of their competitors who have not.

Can we take this logic to the next stage? If our problem is that our institutions are failing to balance the power of the market, can the private sector help strengthen these institutions? If they could, should they? These are the questions I take up in the next chapter.

7

PROTECTING WHAT HAS MADE US RICH AND FREE

Markets, Politics, and the Future of the Capitalist System

> If men were angels, no government would be necessary. If angels were to govern men, neither external nor internal controls on government would be necessary. In framing a government which is to be administered by men over men, the great difficulty lies in this: you must first enable the government to control the governed; and in the next place oblige it to control itself.
>
> —JAMES MADISON, *FEDERALIST PAPERS, NOS. 10 AND 51*

In the end, the issues central to reimagining capitalism can only be addressed by *limiting* the power of business. But our wholehearted embrace of shareholder value maximization at almost any cost—and the systematic devaluing of government that has followed in its wake—means that in many countries national institutions are not currently well equipped to hold markets in check. The media are under sustained attack, and the very idea of democracy is falling out of fashion.[1] Moreover, as noted earlier, many of the problems we face require global solutions, and we have as yet only a very preliminary sense of what globally inclusive institutions might look like. An enormous amount of power is concentrated in the private sector at a time when national institutions are under stress and our global institutions remain relatively weak. What can be done?

THE KEY TO prosperity for both business and society at large is to understand free markets and free politics as complements rather than as adversaries. Free markets need democratic, transparent government if they are to survive—as well as the other institutions of an open, inclusive society including the rule of law, shared respect for the truth, and a commitment to a vigorous free media. Similarly, free governments need free markets. Without the growth and opportunity that truly free and fair markets provide, many societies have trouble maintaining their legitimacy or upholding the minority rights that are at the heart of effective democratic governance.

Reimagining capitalism through the push to create shared value, rewire finance, and find new ways to cooperate will make an enormous difference, but on their own they are not enough to build a just and sustainable society. Effective government action is the missing piece, but the choice is not between markets and government. Genuinely free and fair markets cannot survive without government. The choice is between *inclusion*—transparent, democratic, effective, market-friendly government supported by a strong society and a free media—and *extraction*, the rule by the few on behalf of the few. Free markets need free politics. It is time for the private sector to play an active role in supporting them.

Both environmental degradation and inequality are systemic problems that cannot be solved without government action. Arresting climate change requires decarbonizing the world's energy supply, radically upgrading the world's buildings, changing the way we build cities, remaking the world's transportation networks, and completely rebuilding agriculture. These are massive public goods problems that not even the most sophisticated self-regulation can solve. We need governments to provide either the economic incentives that will move firms to action, or the regulations that will force everyone to do the right thing. Business, in its own interest, must take the lead. Without good government and free politics, the free market will not survive.

Energy demand is projected to double over the next fifty years.[2] Stopping global warming means ensuring that every new plant that's built is carbon-free. It also means shutting down or decarbonizing the world's *existing* fossil fuel infrastructure. These are tasks that only government action—whether it's in the form of a carbon tax or simple regulation—can achieve. Business was able to make real progress in slowing the deforestation of the Amazon—but only with government help. Now that the Brazilian government has changed its policies, rates of deforestation have skyrocketed.[3] The businessmen who built the White City were only able to curb Chicago's pollution as long as they could use the threat of legal sanction to shut down polluters. Once they lost political support and juries refused to convict, the pollution returned.[4]

Inequality presents a similarly tough set of deeply intertwined systemic problems that can only be fully addressed through government action. Providing every child with the education and the health care that he or she needs to be able to compete in the modern economy is clearly table stakes—but table stakes that can only be effectively provided by the state. Moreover, they are not enough to ensure real equality of opportunity. Only about 20 percent of any given student's success is a function of his or her education, while about 60 percent is attributable to family circumstances—and most particularly, to the family's income.[5] Children growing up without adequate nutrition or adequate childcare, and with parents who are working too hard and under too much stress to be able to support them in their schoolwork, are much less likely to succeed.[6] Only government can address the structural factors that drive inequality and raise incomes at the bottom of the income distribution.

Between 1946 and 1980, US total pretax national income nearly doubled. The poorest half of the population saw their income slightly more than double, while the richest 10 percent saw a slightly smaller increase.[7] Between 1980 and 2014 pretax national income grew by 61 percent.[8] But the income of the poorest half grew by only 1 percent, while the income of the top 10 percent grew by 121 percent, and the income of the top 1 percent more than tripled. Average CEO pay—which was about 30 times average worker compensation in 1978—was 312 times average compensation in 2017.[9] Just over half of today's public school students qualify for free or reduced-price school lunches—a classic proxy for poverty.[10]

We cannot provide real opportunity until and unless we can raise wages. Many firms believe—however incorrectly—that they simply cannot afford to raise wages. Recall that Walmart's stock price dropped 10 percent on the day the firm announced it was spending roughly $3 billion to raise its minimum wage by approximately $2.50/hour.[11] Raising it another 50 percent to $15/hour would cost billions more. In 2018 Walmart made about $20 billion in operating profits. That may sound like a lot, but it's only about 4 percent of sales, and increasing labor costs by billions of dollars without increasing sales or productivity could easily trigger a wholesale flight from the stock.[12] It may well be that if the firm could adopt the kinds of labor practices that have allowed Costco or Mercadona to pay over the odds, Walmart too could raise wages.

But such a transformation would be hugely disruptive—and there are many firms that continue to believe that they cannot give their workers a decent wage until and unless all of their competitors are forced to do the same thing.

Moreover simply increasing spending on education and unilaterally driving up wages is unlikely to reduce inequality substantially without moves to address the full range of factors that drive inequality in the first place, from uncontrolled globalization and the decline of organized labor, to changes in the tax code that favor the rich, to the increasing concentration in many industries and the failure to invest in infrastructure. These are all issues that can only be addressed through political action.

OF COURSE, THE idea that business could play a central role in strengthening existing or creating new inclusive institutions might seem, at first glance, a little far-fetched. The very idea of government has been under assault for decades. Ronald Reagan, for example, famously declared in his inaugural address as president of the United States that "in this present crisis government is not the solution; government is the problem."[13] Grover Norquist, the influential head of Americans for Tax Reform quipped in an interview, "I don't want to abolish government. I simply want to reduce it to the size where I can drag it into the bathroom and drown it in the bathtub."[14]

Trust in government—and in the idea that governments can be relied on to fix society's problems—is at an all-time low.[15] But these perceptions are a function of a systematic campaign to discredit government, not of the role that governments could—and have—played in building just and sustainable societies.

One of the most important roots of the triumph of the "free market at almost any cost" ideas that characterized the United States in the 1980s and 1990s was an intellectual and cultural movement that was bankrolled by the private sector. Much of the funding for the Mont Pelerin Society, an international group of scholars that included conservative economists like Friedrich Hayek and Milton Freidman and met regularly for some years to develop a rigorous academic basis for their ultra-free market ideas, came from the business community.[16]

In the decades following World War II, businesspeople funded a number of radio programs and popular magazines that featured the ideas of Ludwig von Mises, Friedrich Hayek, and other neo-liberal thinkers. For example, Howard Pew, president of Sun Oil, funded James Fifield's radio program, the *Freedom Story*, and Billy Graham's magazine, *Christianity Today*. These platforms combined free-market Hayekian concepts with broader social and moral themes, creating a network that supported conservative activism. Resources were also channeled to libertarian think tanks such as the American Enterprise Institute in an attempt to communicate free market, anti-government ideas to policy makers and journalists.

Wealthy business leaders committed to free market thinking also made a concerted effort to influence academic opinion. For example the John M. Olin Foundation, founded by the eponymous industrialist, spent hundreds of millions of dollars between 1960 and 2005 to develop and disseminate the expansion of law and economics as a legal discipline, underwriting the costs of a majority of its early programs and fellowships. A director of the Olin Foundation explained that while law and economics seemed neutral, it possessed "a philosophical thrust in the direction of free markets and limited government," and that it was a way to fund conservative legal scholarship without drawing backlash from university deans. Leading schools such as Harvard Law School and Columbia Law School received substantial funding for new law and economics programs in the hope that they would influence other schools.

Charles and David Koch, sole owners of Koch Industries and two of the richest men in America, were—until David Koch's passing—the de facto leaders of the continuing effort to reduce the size and power of the US government. That mantle now rests on Charles's shoulders alone. Throughout the 1980s and 1990s, the brothers funded a variety of organizations opposing efforts like environmental regulation, cap and trade legislation, and health care reform. Beginning in 2003 they also began to convene twice-yearly donor "seminars" at which wealthy individuals—largely business leaders—were exposed to ultra-free market ideas as well as to practical political strategies for implementing them. By 2010 more than two hundred wealthy donors were regular attendees. The network is committed to cutting taxes, blocking or eliminating business regulation, reducing funding for public education and social welfare initiatives, undercutting public and private labor unions, restricting easy voter registration, and cutting back voting days and hours. It continues to thrive and routinely funds investments designed to generate ideas, research, and change in higher education. But its most extensive efforts have been devoted to the creation of Americans for Prosperity, "a general-purpose federation of organizations." The members of the federation invest in advertising, lobbying, and grassroots agitation. By 2015 the federation had a budget of $150 million and five hundred staff. In 2015, 76 percent of the new political organizations on the right were affiliated with the Koch network, while 82 percent of new extraparty funding was flowing through Koch-affiliated consortia.[17]

THE BELIEF THAT government is actively destructive—that it means unresponsive bureaucrats, high taxes, and endless regulation—has thus been at least partially constructed by a more than fifty-year campaign. These perceptions are an artifact of our present moment, not of the role that government could—and has—played in building just and sustainable societies. Two charges are laid against governments: that they tend toward tyranny and particularly that they seek to replace the free market with state control or central planning, and that they are hopelessly gridlocked and ineffective. Some governments are indeed tyrannical, and some governments are broken. But they have not all always been so, some are not so now, and none have to be so going forward.

Building the Governments We Need: The Big Picture

The question of what kind of political system best supports economic growth and social well-being is, of course, highly controversial. In the 1980s and 90s political thinking in both the developed and developing world focused on the role of free markets in driving economic prosperity and political freedom. Global economic development was guided largely by the "Washington Consensus," a view of the world that focused overwhelmingly on the power of free markets to drive growth. The Consensus led influential bodies such as the World Bank and the International Monetary Fund (IMF) to push developing countries to enact far-reaching deregulation and privatization, to open domestic markets to global trade, and to permit free capital flows as roots of development—all without explicit attention to the health of local political or social institutions.

It's now clear that this was a mistake.

Empirically, many of the states that implemented the Washington Consensus failed to do as well as expected. In post-Soviet Russia in particular, the rapid liberalization of markets was followed by a descent into an extreme form of crony capitalism, while the so-called Asian tigers—especially Taiwan, Singapore, and the Republic of Korea—found economic success by pairing the development of their own markets with heavy government intervention. A 2000 study that found that differences in political and social institutions explained about three-quarters of per capita differences in income across formerly colonized nations provoked a flood of further research.[18] This work eventually confirmed what historians and political scientists had never stopped saying—namely that while economic growth and social well-being are often enormously advanced by the presence of free markets, they are also critically dependent on a host of complementary institutions.

A remarkably coherent consensus has emerged about the fundamental pillars on which any successful system must be based. Scholars now differentiate between "open access" regimes based on "inclusive" institutions—regimes like those in Germany, Chile, the Republic of Korea, and the United States—and "closed" regimes based on "extractive" institutions such as those in Russia, Venezuela, Angola, the Democratic People's Republic of Korea, and Turkmenistan.

The distinction between inclusive and extractive institutions was first emphasized by Daron Acemoglu and James Robinson in their book *Why Nations Fail*. They defined inclusive economic institutions as those that support the effective functioning of a free market, and inclusive political institutions as those that enable the public to participate in the political process and monitor the government. In contrast, extractive institutions concentrate both political and economic power in the hands of the elite.

Extractive regimes are tyrannies. In an extractive society the political and economic power is concentrated in a small elite. The rule of law is only sporadically enforced, the media is a tool of the state, the rights of minorities are routinely abused, and the right to vote—if it is exists—is systematically manipulated and controlled. Free markets rarely thrive under extraction, since the cadres of the elite control the law, and usually use it to create systematic advantage for themselves and their friends, pushing the society toward crony capitalism. The earliest forms of organized society were extractive. In ancient Egypt and feudal Europe, political power was limited to the few who controlled military force.[19] They appropriated all the available economic surpluses for themselves, which in turn enabled them to finance and maintain a system of political and economic control.[20]

Examples of Economic and Political Institutions, Inclusive or Extractive

	ECONOMIC	POLITICAL
Inclusive	Secure property rights Effective education & job training systems Open markets with low entry costs Balanced, fair employer-labor relations Consumer protections Environmental regulation Antitrust enforcement	Democratic pluralism Voting rights Checks & balances in government Free media Freedom of speech & other personal rights An impartial judiciary Protection for minority rights
Extractive	Weak property rights Crony capitalism Widespread anticompetitive monopolies Forced or extractive labor Disregard of externalities	Monarchy / oligarchy / single-party rule System of elites or nobility Suppression of freedoms of expression Patronage networks Influential but opaque interest groups

Adapted from Daron Acemoglu and James A. Robinson, *Why Nations Fail: The Origins of Power, Prosperity, and Poverty* (New York: Crown Books, 2012).

Inclusive regimes are open, democratic, and accountable. They allow anyone—no matter who their parents are—to participate in political and economic life. They are characterized by two core institutions. The first is participatory government. The second is the free market. As I suggested above, the two are complements, and need each other to survive. Both are fragile. Governments continuously seek more power, more wealth, and more control—while markets similarly seek constantly to undermine the rules that constrain them, to seek less regulation, lower taxes, and more power. They need each other—and the other institutions of a free society if they are to stay in balance: the impartial rule of law; a voice for labor; the preservation of minority rights; a free and effective press; and a vigorous, open, and effective democracy.

Where do inclusive institutions come from? Free markets and free political institutions first began to flourish on a large scale in Europe. In some cases, an emerging merchant class pushed extractive rulers to share power. In others, the potential for political inclusion to stimulate economic growth and to drive trade led governments to share power as a way of increasing their ability to thrive in the face of military threats.[21] For example, in medieval Venice, a contractual institution called the *colleganza*—a precursor to joint stock companies—enabled wealthy financiers to provide capital to traveling merchants for long-distance trade. Profits from the challenging journey were shared, providing traders, who were chosen on the basis of merit rather than social standing, with the potential to accumulate significant wealth. In time, this economic inclusion drove political inclusion, as the growing merchant class placed constraints on Venice's ruler (the Doge) by eliminating hereditary rule and creating a parliament. Venice thrived from the tenth to thirteenth century as a result, until the *Serrata* or "closure" in the early years of the fourteenth century, when an exclusive group of wealthy merchants succeeded in limiting access to the *colleganza* and to

Holding the Balance Between Governments and Markets

Free Government

Rule of Law
Free Press
Respect for Minority Rights
Real Democracy
A Voice for Labor

Free Market

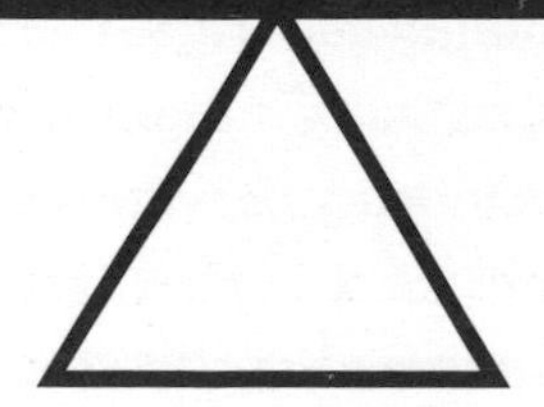

the parliament. The change enabled these families to dominate the Venetian economy and its politics for the next two hundred years—and marked the beginning of Venice's long decline.[22]

The emergence of inclusive institutions in seventeenth- and eighteenth-century Britain was another important turning point. In the English Civil War of 1642–1649 and the Glorious Revolution of 1688–1689, a substantial middle- to upper-class bourgeoisie, many of whom had made their money in trade, played an important role in ushering in democratic reforms.[23] During the Civil War, they executed the king and brought the country under parliamentary rule. The monarchy was restored in 1660, but with significantly restricted powers, and following the Glorious Revolution, an alliance of commercial interests and aristocrats forced the adoption of constitutional guarantees that further checked the monarchy's power and protected elections and free speech. Both the American (1775–1783) and French Revolutions (1789–1799) can be similarly interpreted as a struggle between a growing commercial class and a traditional monarchy. In each case these political revolutions were followed by commercial revolutions—by the breaking of aristocratic and monarchical control over the economy in favor of economic competition open to (almost) all.

The history of the United States is a classic example of the power of inclusive institutions. Following the Revolutionary War, the United States established unprecedented checks and balances on political power, fundamental rights for citizens, and free elections. The political and economic mobility supported by these institutions meant that in principle any common (white) man could acquire economic power—a radical notion in the eighteenth century—and this mobility laid the foundations for the enormous dynamism of the United States' nineteenth-century economy. The shared belief that (provided you were white and male) your potential was limited only by your intelligence and your hard work, coupled with the widespread availability of cheap land and the absence of an entrenched ruling elite meant that the United States saw quite extraordinary levels of social mobility.

In the nineteenth and twentieth centuries, this institutional regime was further strengthened by the broadening of the franchise, public funding of education, and the development of institutions such as a vigorous free media and a wide range of labor, social welfare, consumer protections, and antitrust legislation. Partly as a result, the United States has historically had not only an extraordinarily innovative and dynamic economy but also among the highest levels of social welfare of any developed nation.

Under effective inclusive institutions, effective governments are valued partners in sustaining both the free market and a free society. The US Department of Defense, for example, was the first customer for the computer industry—an industry that got its start using discoveries that had been funded by federal funds. Massive state investments in research and development led to the revolutionary technologies that underpin the iPhone and the iPad, the internet, GPS, touch-screen displays, and most communication technologies. Agricultural extension schools funded by the federal government were instrumental in diffusing the best practice techniques that helped to make US agriculture the most productive in the world. Government money built the roads, ports, and bridges that sustain the economy.[24]

Government regulation has solved a wide variety of environmental problems. In 1973, for example, the chemists Frank Sherwood Rowland and Mario Molina discovered that the chlorofluorocarbon (CFC) molecules used as aerosols and refrigerants were stable enough to reach the stratosphere, and that their presence there would cause the breakdown of the ozone layer that protects life on Earth from the sun's ultraviolet radiation. High levels of ultraviolet rays cause skin cancer in humans and significant damage in other animal and plant life. Rowland and Molina recommended that CFCs should be banned as soon as possible.[25]

This idea was strongly contested by the CFC industry, which at the time had at least $8 billion in sales and employed over six hundred thousand people. The chair of the board of DuPont was quoted as saying that the ozone depletion theory is "a science fiction tale . . . a load of rubbish . . . utter nonsense."[26] DuPont, the largest CFC manufacturer, speculated publicly that the costs of phasing out CFCs could exceed $135 billion in the United States alone, and that "entire industries could fold."[27]

But twelve years later, three scientists discovered a hole in the ozone layer over Antarctica that was much larger than anyone had expected. One estimate suggested that if the question of CFCs was not addressed, by 2030 an additional six hundred thousand people would die of skin cancer and an additional eight million people would develop cataracts. There would also be significant damage to plant and animal life. Despite continuing opposition, the Montreal Protocol—an international agreement to phase out ozone-destroying chemicals—was approved a year later to address the threat. The protocol has been remarkably successful. It proved possible to find CFC substitutes relatively quickly, and the Antarctic ozone hole is expected to return to its 1980 status by 2030. It has also reduced global GHG emissions by about 5.5 percent.[28]

In the United States, the passage of the Clean Air Act in 1990 was similarly effective. For example, the program to control acid rain—the SO_2 (sulfur dioxide) allowance trading program—cost no more than $2 billion a year and led to benefits of between $50 and $100 billion a year in reduced mortality.[29]

Government regulations keep food and water supplies safe—and ensure that workers are not routinely abused on the job. The adoption of national pension schemes and government-funded health care for the elderly has meant that millions of people no longer face the prospect of hunger and sickness in their old age. No regulation and no government program is perfect, of course, and government regulators can sometimes be difficult. But this is inherent in the nature of the institution—the give and take of the political process in combination with a focus on the public good rather than on private profit will always mean that governments look less "efficient" than the private sector. But efficiency is not the right criterion. The right criteria are whether the government is clean, responsive, transparent, and democratic.

Economic growth in societies with strong inclusive institutions is more consistent, and inclusive societies are significantly more prosperous than societies living under extraction. Inclusive institutions are also a strong determinant of individual well-being. Inclusive societies are happier and longer-lived. They have lower-income inequality, greater socioeconomic mobility, and greater social freedoms.[30] While per capita GDP is a strong determinant of life satisfaction for poor countries, when GDP is greater than about $15,000/year, individual happiness is correlated not with income but with the presence of inclusive political institutions.

In short, democratic governments and the other institutions of a free society are a primary source of economic growth and individual well-being. The problems that confront us now reflect the fact that we must build effective global institutions if we are to address the global problems we face—but across the world inclusive institutions are under sustained attack.

Business must become an active partner in shoring up the inclusive institutions that we have and in building the new ones that we need. This is not a question of supporting specific policies or of pushing a particular set of political values. This is about supporting the foundations of our society. Business has to learn to think systematically. The question should not be "Would this particular policy benefit me," but "How do we protect the institutions that have made us rich and free?"

Many firms are already engaged with their local communities (as we saw earlier with the example of Minneapolis-St. Paul), working with local government to build the public goods that every society needs. Efforts have to be expanded to both national and global levels and focused on three key issues. The first is around minority rights and inclusion. Business should do what it can to ensure that everyone in every society—regardless of race, gender, or ethnicity—has the opportunity to be a full member of that society. The second is the need to price or regulate major environmental externalities. Free markets only work their magic when everything is properly priced. As long as firms can burn fossil fuels, poison the oceans, and discard their waste without penalty, they will continue to drive global warming and the destruction of the biosphere. Business must press for the legislation that would force every firm to behave "well." Lastly and perhaps most importantly, business must do everything it can to preserve and strengthen democracy and civil society.

Advocating for Minority Rights

Respect for the rights of minorities is one of the fundamental pillars of an inclusive society, and a key indicator of the existence of healthy institutions. Building a just and sustainable society requires not only the protection of property rights and political rights but also civil rights, or equality before the law. No society can be inclusive if it discriminates between groups in the provision of public goods such as justice, security, education, and health. Business is immensely powerful and beginning to flex its muscles in opposition to discrimination. The fight to protect LGBTQ rights is one example of what this might look like to practice.

In the United States views of LGBTQ people have changed enormously in the last twenty years: 70 percent of the population—including majorities in both political parties—now say that "homosexuality should be accepted," up from 46 percent in 1994.[31] Moreover, 61 percent favor same-sex marriage, up from 38 percent in 2002, and a (very slim) majority believe that trans people should have civil rights.[32]

While the bulk of the heavy lifting on this front has been driven by the enormous courage and persistence of LGBTQ people themselves, the movement has been assisted by many of the large global corporations, who were early supporters of gay rights. AT&T adopted a policy prohibiting discrimination against employees based on sexual orientation in 1975, and IBM included sexual orientation as part of its global nondiscrimination policy in 1984.[33] In 1992 the Lotus Development Corporation became the first publicly traded company to offer benefits to its gay employees.[34] IBM extended health care coverage to same-sex couples four years later, while Walmart began to provide company-wide health insurance benefits for all the domestic partners of its workers in 2013. Among the Fortune 500, 85 percent include gender identity protections in their nondiscrimination policies (up from 3 percent in 2002).[35] Sixty-two percent offer transgender-inclusive health care coverage, up from zero in 2002.[36] When the Corporate Equality Index, which rates major companies on pro-LGBT policies was first introduced in 2002, only 13 companies achieved perfect scores, of the 319 surveyed. Today, even though the index has been revised to make it more stringent, 366 of 781 businesses score a 100 percent, including fourteen of the top twenty on *Fortune*'s rankings of the largest companies in the United States.

But the battle has not been won. About half of the estimated 8.1 million LGBTQ workers aged sixteen and older live in states without legal protections against sexual orientation and/or gender identity employment discrimination, and the number of hate crimes against LGBTQ people has not fallen significantly.[37] More recently a number of cities and states have passed legislation that appears to be designed to make it legal to discriminate against LGBTQ people. For example, in March 2015 Mike Pence, then governor of Indiana, signed the Religious Freedom Restoration Act (RFRA). The act was signed in a private ceremony attended by several groups that had been publicly opposed to gay marriage, and critics charged that the legislation would allow organizations to use religion to legally justify discrimination on the basis of a person's sexual orientation.

Many CEOs—both within Indiana and beyond—reached out publicly in an attempt to reverse the legislation. Immediately before the act was passed, a number of Indiana-based technology company CEOs, including the CEOs of Clear Software, Salesforce, CloudOne, and Salesvue sent a letter to Mr. Pence urging him to veto the measure, saying:

> As leaders of technology companies, we not only disagree with this legislation on a personal level, but (believe that) the RFRA will adversely impact our ability to recruit and retain the best and the brightest talent in the technology sector. Technology professionals are by their nature very progressive, and backward-looking legislation such as the RFRA will make the state of Indiana a less appealing place to live and work.[38]

Following the signing of the act, Tim Cook, the CEO of Apple (the first CEO of a Fortune 500 company to come out as gay) tweeted that he was "deeply disappointed" in the law. Yelp CEO Jeremy Stoppelman remarked,

> [It] is unconscionable to imagine that Yelp would create, maintain, or expand a significant business presence in any state that encouraged discrimination by businesses against our employees, or consumers at large. . . . These laws set a terrible precedent that will likely harm the broader economic health of the states where they have been adopted, the businesses currently operating in those

states and, most importantly, the consumers who could be victimized under these laws.[39]

The CEOs of Anthem Inc., Eli Lilly and Company, Cummins, Emmis Communications, Roche Diagnostics, Indiana University Health, and Dow AgroSciences—all companies with significant operations in Indiana—called on the local Republican leadership to pass legislation to prevent "discrimination based upon sexual orientation or gender identity." Bill Osterle, the CEO of Angie's List, a home-services website that had recently announced that it would expand its operations in Indiana, announced that the firm would postpone the $40 million expansion, putting as many as one thousand new jobs at risk.

A week later the legislature passed an amendment to the bill, clarifying that it could not be used to defend discrimination against LGBTQ people.[40] A month later the governor of Arkansas also signed a revised "Religious Freedom" law following public pressure from the CEO of Walmart, who publicly requested that the governor veto the original version of the bill on the grounds that it legitimized discrimination against LGBTQ people.

A similar response greeted North Carolina's legislature when it passed the Public Facilities Privacy and Security Act, more commonly known as the "HB2" or the "bathroom bill" in March 2016. The bill eliminated a local city ordinance that made it illegal for businesses to deny service to LGBTQ individuals and would also have allowed transgender people to use restrooms that corresponded to their gender identity. Under the state bill, transgender people were required to use public restrooms corresponding to the gender listed on their birth certificates.

The day after the bill passed, a number of companies—including American Airlines, Red Hat, Facebook, Apple, and Google—issued statements opposing HB2. A few days later more than one hundred other CEOs and business leaders signed a letter expressing their concerns about the bill. The cofounder of PayPal, Max Levchin, who the previous year had told CNN that opposing the Indiana law was "a basic human decency issue," canceled plans to open a new operations center in Charlotte, the largest city in North Carolina and the source of the original ordinance, costing the state as many as four hundred new jobs. A week later Deutsche Bank announced that it was canceling its plans to create 250 jobs in Cary, North Carolina. A year later, North Carolina passed a partial repeal of HB2.

Taking these kinds of aggressive public positions is not always easy or cheap. Walmart's employees and customers are almost certainly deeply divided over the question of LGBTQ rights. Osterle, the CEO of Angie's List, was widely attacked. The president of one local conservative group said, "I see what he did as . . . nothing short of economic terrorism." The One Million Moms blog labeled the firm "a bully, plain and simple" and called for a boycott. Dan Schulman, the CEO of PayPal, recalled, "We got many accolades from obviously lots of different people for [speaking out], but we also got a lot of people who disagreed with that decision. I got a lot of threats, personal threats."

Discrimination is a bread and butter issue for the world's largest corporations since their millennial employees are passionate about the issue and demand that their employers take a stand against it. But opposing discrimination—whether it's on the basis of gender, race, or ethnicity—is also a core moral value for many business leaders.

On August 14, 2017, for example, Ken Frazier, the CEO of the US pharmaceutical giant Merck, announced that he was resigning from President Trump's Manufacturing Council. He was reacting to the president's suggestion—following violent clashes in Charlottesville, Virginia, where a rally organized by white supremacists had led to the death of a young woman—that there was "blame on both sides." Frazier issued a statement that said, in part,

> Our country's strength stems from its diversity, and the contributions made by men and women of different faiths, races, sexual orientations and political beliefs. America's leaders must honor our fundamental values by clearly rejecting expressions of hatred, bigotry, and group supremacy, which run counter to the American ideal that all people are created equal. As CEO of Merck, and as a matter of personal conscience, I feel a responsibility to take a stand against intolerance and extremism.

Talking to the press about the incident a year later, Frazier—who is African American and whose grandfather was born into slavery—recalled: "It was my view that to not take a stand on this would be viewed as a tacit endorsement of what had happened and what was said. I think words have consequences, and I think actions have consequences. I just felt that as a matter of my own personal conscience, I could not remain."[41]

Given Frazier's position as the CEO of a major drug company, his decision was not without risks. President Trump had actively campaigned on the promise to bring down drug prices, and indeed within an hour of Frazier's statement he tweeted, "Now that Ken Frazier of Merck Pharma has resigned from President's Manufacturing Council, he will have more time to LOWER RIPOFF DRUG PRICES!" But within a week, all of the other CEOs on the council had resigned. They were all personally attacked by President Trump.

These may seem like small actions. But they are suggestive of a willingness on the part of some senior business leaders to challenge powerful politicians in the service of strongly held values. Should US politics start to move toward a more inclusive stance on issues of race, gender, and ethnicity, I suspect that the private sector's willingness to engage in these issues will be seen as an important part of the movement that made it possible.

Going Global: The Private Sector and Climate Policy

Business must push governments everywhere to address climate change, insisting that policy be based on current science, and advocating strongly for market-friendly policies that could help us to avert disaster. Appropriate regulation—something like a carbon tax or a carbon cap—would not only allow the global economy to decarbonize at minimal cost but would also open up billions of dollars in new market opportunities. Decarbonization will be expensive. But unchecked climate change will cost billions of dollars more. Current estimates suggest that climate change could cost the US economy as much as 10 percent of GDP by the end of the century and destabilize the world's food supply.[42] The IPCC estimates that keeping GHG emissions to a level that offers a 66 percent chance of not exceeding 2°C warming would cost 3 to 11 percent of world GDP by 2100.[43] But leaving global warming unchecked might cost 23 to 74 percent of global per capita GDP by 2100 in lost agricultural production, health risks, flooded cities, and other major disruptions.[44] Unchecked climate change will also impose irreversible harm on coming generations. Just as the world's business community has committed to global inclusion, so it should commit to ensuring that we leave a healthy planet to our children.

Many in the private sector are already beginning to move in this direction. In British Columbia, private sector support was critical to putting the province's climate tax in place. In the United States it was central to the adoption of the Regional Greenhouse Gas Initiative (RGGI), the Northeast/mid-Atlantic carbon trading system, and to the design and passage of California's commitment to be 100 percent carbon free by 2045.

In April 2019, Gary Herbert, the Republican governor of Utah, signed the Community Renewable Energy Act. The legislation required Rocky Mountain Power, Utah's power provider, to provide 100 percent renewable energy to those communities in the state that requested it—paving the way for towns and cities across the state to move to renewable power. Its passage crowned three years of quiet negotiation between business, those cities that wished to move to renewable power, and the utility.

Bryn Carey is one of the businesspeople who helped build political support for the move. In 2004 he founded Ski Butlers, a ski equipment delivery and rental business, out of a single-car garage in Park City. He quickly became aware that climate change represented a significant threat—not only to his business but to the state and the sport that he loved. Utah is currently one of the five US states warming most rapidly. In the last forty-eight years, the average temperature has increased by over 3°F, and the snow pack has fallen significantly, threatening not only the ski industry but also the state's water supply.[45] In 2012 Carey spent months trying to persuade the business community in Park City to back an initiative to install solar panels on every roof in the city, but without success, ultimately deciding that only politics could drive action. In 2015 he rallied his own employees, local activists, and other residents—"dozens" of people in total—to show up for a city council meeting. The following year the council passed a resolution committing the city to be powered 100 percent by renewable energy by 2032.[46] In July 2016 Salt Lake City passed a similar resolution, and in 2017 a number of other Utah communities followed suit.

But in Utah, cities that wished to switch to renewable power have only two choices—build their own electric utility from scratch, or buy from Rocky Mountain Power, a state-regulated monopoly that was notorious for burning coal. By 2015, Salt Lake City had violated federal air quality standards for over a decade, largely because of the city's reliance on coal-fired electricity. In response Mayor Jackie Biskupski, the city's recently elected mayor, suggested that Rocky Mountain agree to fund the investments that could make Salt Lake City's commitment to renewable energy feasible.

Over the next two years the mayor—joined by a number of other allies from the environmental community, business, and mayors of the other cities that had committed to using renewable power—negotiated quietly with the utility to craft legislation that might make this possible. The coalition was helped by the fact that the economics of renewable energy are changing so fast. In December 2018 PacifiCorp, the entity that owns Rocky Mountain Power, published a report suggesting that thirteen of its twenty-two coal plants were more costly to run than the available alternatives and that shutting them down could save millions of dollars.[47] But none of the company's Utah plants made the list since the utility had not yet fully paid down the debt incurred to build the plants, and the utility demanded compensation if it were to close them down prematurely. The final bill provided for those communities that switch to renewables to continue paying down the coal debt. Reflecting on the complicated negotiations that made the settlement possible, Biskupski recalled, "Did we have moments where people wanted to throw their arms up? Sure we did. But when that happens, you have to bring everybody back to the table and remind them: this is the journey. This is the commitment. The people, they want clean air."[48]

More than two hundred US businesses, cities, and counties are committed to 100 percent clean, renewable energy.[49] "America's Pledge," a nonprofit devoted to keeping track of local commitments across the United States, estimates that these commitments add up to a 17 percent reduction in GHG levels from 2005 levels by 2025. It further suggests that strategies to reduce emissions further that are "high-impact, near-term, and readily available for implementation by local actors" could take this number to 21 percent.[50]

Broader engagement with the goals of the coalition across the US economy could potentially lead to more than a 24 percent reduction below 2005 levels by 2025. That decrease would put America "within striking distance of the Paris Pledge." The report closes by claiming that "decarbonization can be led by the bottom-up efforts of real economy actors . . . but only with deep collaboration and engagement."

In 2017, when Trump declared that he was going to withdraw the United States from the Paris Agreement[51]—joining Syria and Nicaragua as the only countries not committed to taking action against climate change—the CEOs of thirty US companies, including those from Apple, Gap, Google, HP, and Levi Strauss—published an open letter urging him to rethink the decision. Elon Musk, the CEO of Tesla, and Bob Iger, the CEO of Disney, resigned from the President's Advisory Council in protest.[52]

An even more ambitious collaborative effort called "We Are Still In" now "includes 3,500 representatives from all 50 states, spanning large and small businesses, mayors and governors, university presidents, faith leaders, tribal leaders, and cultural institutions." It is committed to catalyzing action at the local level to ensuring that the United States meets its commitments under the Paris Agreement.[53] As of this writing more than two thousand businesses are signatories to the agreement, all of them formally committed to working with national governments and local communities to reduce GHG emissions. The coalition attended the international climate negotiations at COP24 in December 2018, acting as a "shadow delegation" and meeting with national governments and delegates to the conference to make the case for a strong set of rules to operationalize the Paris Agreement.

My colleagues who work in international climate negotiations tell me that this show of support from the private sector is critically important in keeping the international climate negotiations alive—but that our situation remains desperate. The private sector must make addressing climate change its first ask of every government at every opportunity.

Supporting Democracy

Supporting inclusion and pushing hard for appropriate environmental policy are critical tasks. But the most important issue facing business is to prevent the further destruction of our institutions. Our political institutions are under threat almost everywhere. Gerrymandering is supporting increasingly polarized legislatures and furious partisan battles. Politicians craft rules to restrict turnout and assault the free press. Judicial independence is increasingly being compromised. More and more money is rushing into politics, creating the perception—if not always the reality—that politician are for sale. If our political institutions are to be genuinely free and fair, everyone's voice must be heard, but potential voters are becoming increasingly angry and cynical.

These are dangerous developments for business. As I said above, the alternative to strong, democratically controlled government is not the free market triumphant. The alternative to democratically controlled government is extraction—the rule by the very few for the very few. Extractive elites are not fans of the free market. They cannot resist the temptation to write the rules in their own favor, to shut down innovation, and to suppress dissent. They let infrastructure rot, underinvesting in roads, R&D, hospitals, and schools. If democracy dies, so—in the end—will liberty, the free market, and the prosperity it brings.

Business must demand that the rules of the game are determined democratically. This means actively supporting measures that make it easier for people to vote. Voting rates in the United States, for example, are among the lowest in the world. In the 2014 midterm elections only 33 percent of the voting age population actually voted—the lowest turnout in any national election of any advanced democracy (except Andorra) since 1945.[54] It also means pushing back against any effort to suppress voting rights. In November 2018, for example, a Florida ballot initiative to restore voting rights to convicted felons passed by nearly 65 percent. But in May 2019 the Florida legislature passed legislation insisting that felons could only vote if they had paid off all their court-ordered fines, effectively nullifying the intent of the ballot initiative. This should be a redline—business should call out and actively resist these kinds of measures.

It means collaborating with those who are trying to reduce the amount of money in politics. In the United States spending on lobbying more than doubled between 2000 and 2010 (from $1.57 billion to $3.52 billion) and has since stabilized at around $3.25 billion per year.[55] Following the Supreme Court's 2010 *Citizens United* decision, external spending on presidential elections surged from $338 million in 2008 to $1.4 billion in 2016.[56] This spending excludes politically motivated donations by the tax-exempt charitable foundations of US companies, which a recent study estimated at $1.6 billion in 2014.[57] Although much of the growth in political spending has likely come from very wealthy individuals rather than from business corporations, the fact that there is now a great deal more corporate money in politics is beyond doubt.[58] This flood of spending might benefit individual firms, but it opens the private sector up to charges of corruption and greatly reduces trust in the democratic process. It must be resisted.

How much of this resistance is already underway? My sense—and I have been actively hunting for it—is not much. A campaign called "Time to Vote" has gained the support of three hundred companies across the country, including Walmart, Tyson Foods, and PayPal, all of which are committed to increasing voter participation through programs such as paid time off to vote and an election day without meetings, and support for mail-in ballots and early voting.[59] Corley Kenna, director of global communications and public relations at Patagonia, one of the firms involved in the initiative, told CNBC, "This campaign is nonpartisan, and it's not political. . . . This is about supporting democracy, not supporting candidates or issues."[60] A group of businesspeople spearheaded by Reid Hoffman, an early employee at PayPal and one of the cofounders of LinkedIn, has put hundreds of millions of dollars into efforts to boost voter turnout and bring new candidates into politics.[61]

These are encouraging signs, but they are nothing like the collective action that will be required to support the inclusive institutions that already exist and to create the new ones that we need. When I'm working with businesspeople, this is always the moment at which nearly everyone gets seriously jumpy. They want to know if something like this has ever happened before. Has business ever been able to rebuild political systems in ways that made them more inclusive? It has indeed.

Building Inclusive Political Systems

There have been several moments when business has played a pivotal role in helping to build inclusive societies. Below I briefly describe three, looking at Germany immediately after the first and second world wars, Denmark in the late nineteenth century, and Mauritius in the 1960s. In Germany business faced a system under such stress there were serious doubts as to whether it could survive. But business built a new way of working with its employees, and together they forged a system that has made Germany one of the world's most prosperous and successful societies. In Denmark an embattled ruling elite helped to build a system in which labor, business, and government have worked together to turn one of the smallest and poorest places in Europe into one of the richest and most equal—and one of the most market-friendly. Lest you write off the successes of Germany and Denmark as a function of their European heritage or their fundamentally homogeneous societies,

I then turn to the history of Mauritius—a society badly fractured along racial lines that was able to build a thriving, multicultural community and a strong free market—that is now one of the most successful countries in Africa. In each case visionary business leaders had the courage and the imagination to try something new: to commit to collaboration in the service of building a society that worked for everyone. In retrospect their decisions look obvious and even easy—but at the time they were not obvious at all.

Germany Builds Business-Labor Cooperation

Following Germany's defeat in World War I, political and economic chaos led to the abdication of Emperor Wilhelm II and the collapse of the German Empire. "Councils" modeled after those of the Soviet Union gained widespread political control, and a number of socialist parties gained increasing prominence. Facing what they believed was a real threat of large-scale expropriation and economic disaster, a group of prominent businessmen reached out to the labor unions in an attempt to restore stability.

Hugo Stinnes, the wealthiest man in Germany, played a particularly important role in this attempt. Stinnes had vast holdings in coal, iron, steel, shipping, newspapers, and banks; one analyst described him as the Warren Buffett of his time.[62] Together with a number of other private sector leaders, he approached moderate union leaders with proposals for a new economic order. In November 1918, the two groups signed the Stinnes-Legien Agreement, which established the eight-hour workday, the recognition of labor unions, a right to establish works councils, and the adoption of sectoral collective bargaining.[63] Stinnes also laid the foundation for a national system of employer representation. Talks between the major firms began in 1918 and were concluded in 1919 when employer representation was unified under the Reich Association of German Industry (RdI). The RdI was organized along industry rather than regional lines, a structure that tended to favor big business interests, and the new association allowed Stinnes and his allies to expand the influence of the Stinnes-Legien Agreement across a broad swath of additional firms.[64]

Between 1933 and 1945, fascism, economic and political turmoil, and the disaster of World War II led to the breakdown of these agreements as the Nazis dissolved both employers' associations and unions. Leading German businessmen despised the Nazis, but chose to collaborate with the new regime, reaping rich profits in the short term but acquiescing in a process that ultimately destroyed both the country and their own wealth. World War II left Germany—and German business—in ruins. More than seven million Germans had died, more than eight percent of the population. Twenty percent of the housing stock had been destroyed. Industrial output and agricultural production was only about a third of what it had been before the war.[65]

Fearing widespread radical unrest, the leading employers joined forces with labor once more. The Federation of German industries (BDI) emerged as the largest employer association, focusing largely on economic advocacy, while the Federation of German Employers' Association (BDA) was formed to manage labor relations. The Confederation of German Trade Unions (DGB) was founded as an umbrella organization to represent the trade unions. The three organizations worked together to revive the tradition of employer-labor relations and collective bargaining that had been established between the wars. All three organizations exist today.[66]

One of the most critical tasks of postwar reconstruction was the revival and standardization of the apprenticeship system. Prior to World War II, the system had not been standardized by either sector or region. There were different certification systems for different skills, and there was considerable variation in the type and quality of the training. Following the war, a national coordinating body spearheaded by several large industrial firms and sponsored by BDI and BDA worked to catalogue the skilled trades, to create and disseminate training materials, and to administer certification exams. All apprentices in the same vocation received the same basic training and certification, and mechanisms were developed to try to ensure that the training was continuously refreshed to respond to new technological developments. The number of apprenticeship programs increased dramatically, and legislation passed in the mid-1950s formally recognized the system.[67]

Business associations and labor unions agreed to bargain annually to set pay and working conditions. These annual agreements are legally binding and now cover about 57 percent of all Germany employees. Most German employers also invest in training; provide childcare support; and provide space, materials, and time for employee-run works councils. Publicly traded German companies above a minimal size are required to include employee representatives on their boards in a system that has become known as "codetermination."

Germany now has one of the world's strongest and most equal economies. In 2017 it had one of the world's highest levels of GDP per head, lower only than Ireland, Norway, Sweden, and the United States.[68] Income mobility—a measure of the degree to which someone born to poor parents has as good a chance of making as much money as someone born to rich parents—is lower than in the Scandinavian countries, but is higher than in the United States, the United Kingdom, France, Japan, or China. Average wage levels are among the highest in the world, and unemployment rates (just under 5 percent as of this writing) are among the lowest.

Despite these high wages, German firms are enormously successful exporters. In 2017 Germany exported $1.3 trillion worth of goods, nearly half the value of total economic output.[69] (In the same year the United States exported only $1.4 trillion, 12 percent of output, while China exported $2.3 trillion, 19.76 percent of output.)

By some measures it is ranked as the world's most innovative economy.[70] Nearly 25 percent of German GDP is in manufacturing (in the United States, for comparison, manufacturing makes up only about 15 percent of output). The World Bank's 2016 Logistics Performance Index ranks Germany's logistics performance and infrastructure as the best in the world.[71]

Eight of the world's one hundred largest companies are German, and the country also boasts a highly successful *Mittelstand*, or group of globally successful small and middle-sized firms. Of the world's roughly 2,700 "hidden champions"—firms that are in the top three in their industry and first on their continent and have less than €5 billion in sales—almost half are in Germany. By one estimate these relatively small firms have created 1.5 million new jobs, grown by 10 percent per year, and registered five times as many patents per employee as large corporations. Nearly all of them survived the great recession of 2008–2009.[72]

Much of this success is driven by German's apprenticeship system. Germany now has one of the most sophisticated apprenticeship training schemes in the world. Students can choose from hundreds of trades for an apprenticeship lasting two to four years, during which they split time between classroom instruction and on-the-job training. Trainers are paid for their time—including the time they spend in class. Training at every company is shaped by standardized occupational profiles, or curricula, developed by the federal government in collaboration with employers, educators, and union representatives, giving employees a standardized qualification that allows them to move between firms.[73]

This story has two major implications for our current predicament. The first is that business may not always recognize where its best interests lie. The disasters of the two world wars forced the German business elite into a relationship with its workforce that many modern American managers would fight tooth and nail, but it was an important contributory factor to a system of institutions that has built one of the most successful—and the most equal—societies in the world. To me this underlines the critical role that purpose-driven firms can play as catalysts for uncovering new ways of working. Just as Nike was blind to the benefits of rethinking its relationship to its supply chain in the 1990s, so many firms remain blind to the critical role that the broader social and political system plays in their success. The second implication is more prosaic. If the private sector is to play a central role in rebuilding institutions, it will need strong employee associations committed to playing a positive and political role in the national conversation.

Fortunately, there are a number of business associations that could play the same kind of role in the national and international conversation that the BDI and BDA played in Germany. In the United Kingdom, the Confederation for British Industry represents over 190,000 firms. The US Chamber of Commerce represents more than three million businesses and in 2017 spent more money lobbying the US Congress than any other single organization.[74] The Business Roundtable could also act as an important focus for action.

Historically neither the US Chamber of Commerce nor the Business Roundtable has taken political positions beyond the immediate interests of its members—and the Business Roundtable's new statement makes no mention of any responsibility for the health of US democracy—but the roundtable was the first broad-based business organization to acknowledge the threat posed by climate change, and the Chamber of Commerce recently adopted a formal position on climate change, acknowledging its reality and calling for the United States to "embrace technology and innovation" and "leverage the power of business" to address it.[75]

Business leaders must insist that we address global warming. This means vigorously supporting the science—more than 70 percent of Americans now say that global warming is "personally important" to them, but 30 percent still deny that it's happening and/or that humans are to blame—and this minority remains politically powerful.[76] Addressing it also requires pushing hard for sensibly designed, business-friendly global controls on greenhouse gas emissions. If the Business Roundtable and the Chamber of Commerce fail to lead this charge, it will be critically important to build alternative associations that can take up the slack.

Globally, the World Economic Forum (WEF) includes "the 1,000 leading companies of the world" and bills itself as the "global platform for public-private cooperation." The WEF is deeply involved in projects designed to create shared value, and is also playing a leading role in a wide range of collaborative/public/private reform efforts. The Global Battery Alliance, for example, is a global platform for collaboration designed to "catalyse and accelerate action toward a socially responsible, environmentally sustainable and innovative battery value chain";[77] the project on strengthening global food systems facilitates multistakeholder dialogues, and the Transformational Leaders Network engages "over 150 action leaders and experts to exchange knowledge, best practices, and experience across regions and sectors."[78] To my knowledge, the WEF has as yet steered clear of anything that might be construed as political action—indeed it has been heavily criticized for its failure to examine the structural factors that have contributed to our current troubles[79]—but it has the membership and the capacity to move in this direction if today's senior business leaders were to decide that doing so is important.

The International Chamber of Commerce (ICC) is less visible but has the potential to play an even more critical role. It is the world's largest business organization, representing more than forty-five million companies in over one hundred countries. Its primary mission is to formulate and enforce the trade rules through which international business is conducted, and it often represents business interests in negotiations with global bodies such as the World Trade Organization, the United Nations, and the G20. The ICC is sponsoring some intriguing efforts designed to improve traceability in the supply chain and has formally committed to the importance of "fully engaging the corporate sector in the implementation of the SDGs [Sustainable Development Goals]."[80] To my knowledge the ICC has as yet steered well clear—at least in public—of anything that might look like explicitly political action; it could, for example, insist that sustainability criteria inform global trade practices, but in principle, the ICC has the connections and the global reach to play a central role in building a more sustainable global system.

It's easy to be cynical about the potential for the private sector to play a role in driving systemic change. Indeed one interpretation of the German experience is that we need the equivalent of a world war to change business attitudes. Fortunately the Danish experience suggests that this is not necessarily the case.

Denmark: Business Responds to National Weakness

Bernie Sanders's championing of Denmark has led many business leaders to roll their eyes. But Denmark is not a socialist country, if by *socialism* is meant state ownership of the means of production. Its economy is a strongly pro-business system in which business, labor, and government work closely together to sustain economic growth—within a structure that was championed by the private sector.

In the second half of the nineteenth century, Denmark was a nation in trauma. In 1864 the country had lost the Second Schleswig War to Prussia and Austria, losing the Duchy of Schleswig and the Duchy of Holstein—territories that had been under some form of Danish control since the twelfth century. This was one of a long line of Danish defeats and left Denmark a small, poor country that could no longer aspire to great power status.

By the 1890s the Danish legislature was divided between the Danish Right Party—an uneasy alliance between big agriculture and Denmark's leading industrialists—and the Social Democrats—the party of the working class. Given the split, the Danish king kept the conservatives in power by filling his cabinet with members of the Danish Right Party. In 1890 the Social Democratic Party had gained a significant number of seats and—fearing (correctly) that they would soon be in the minority—members of the Danish Right Party began to search for other mechanisms through which to maintain their influence.

The creativity of a single business leader proved to be critical in translating this moment of crisis into successful institutional change. In 1896 Niels Anderson, a member of Parliament and a railroad entrepreneur with a particular skill in building consensus, took the lead in forming the Confederation of Danish Employers, or the DA. He sold the idea to his colleagues as a means to influence public policy in the absence of a legislative majority (in 1901 the Social Democrats swept the elections, as he had predicted) and as a means to achieving industrial peace by unifying the voice of business. He was remarkably successful in achieving both goals.

The DA was able to beat back a Social Democratic proposal to create a universal, tax-financed workers' accident insurance system, successfully replacing it with a board composed of government, employer, and labor officials. Accident insurance would not be tax-financed; instead, employers would choose their own private insurance.[81] The success persuaded Danish business—and the Danish government—that the DA could be an important vehicle for business-friendly policy reform.

But Anderson's most important success was in demonstrating that the DA could head off labor unrest—by strengthening the labor movement! In 1897 a wave of strikes hit the metals industries. DA members had already agreed that they would not hire workers from other members of the DA during strikes or lockouts, giving the DA both a united front in its dealing with the unions and persuading a majority of Danish businesses to join the association. Next the association inserted itself as a mediator between employers and labor, actively encouraging labor to become better organized, and asking the unions to refuse to work for firms that did not join the DA. The DA was able to negotiate an end to the strike, and in 1898 the Danish Confederation of Trade Unions, or the LO, was founded with the active assistance and support of the DA.

Two years later a massive three-month labor conflict known as the "Great Lockout" erupted in the steel industry. Together the DA and LO were able to negotiate a "September Compromise," which established a national system of collective bargaining. In 1907 the state gave the DA the power to intervene in sectoral disputes, reinforcing the DA's role as the top employers' association and giving the association a central role in negotiating business labor relations.

Over the next fifty years cooperation between labor, business, and government became increasingly routinized—so much so that by the 1960s Danish conservatives were bragging that they were responsible for a "decisive step in the expansion of the welfare state." In the 1970s and 1980s, economic recession and high unemployment put the Danish welfare state under financial pressure, and the state began to cut benefits and to suspend wage indexation. But in response, employers and labor struck a sequence of bilateral agreements in which employers increased investments in training. In the 1990s the DA played a central role in helping to design a series of "active labor market policies" (ALMPs) that were designed to move unemployed youths into training programs and match them with jobs. The DA worked closely with the LO to build worker support for the ALMPs, and persuaded member firms to participate, facilitating communication between the government and individual firms to ensure their smooth implementation.

Denmark is now one of the world's most successful societies. It has one of the highest average minimum wages in the world—$16.35

in 2015—despite the fact that it does not have minimum wage legislation on the books.[82] GDP per capita is higher than it is in Canada, the United Kingdom, and France. It also has the lowest income inequality among the OECD, with the richest 10 percent of Danes earning only 5.2 times more than the poorest 10 percent.[83] Danes get five to six weeks of vacation a year, and up to a year of paid maternity/paternity leave.[84]

Policy making in Denmark is highly collaborative, bringing government, employers, and unions together in a joint process that goes back over a hundred years. This approach has made possible a unique mixture of relaxed labor regulations (which favor firms) and a strong welfare state (which favors workers), under which firms can fire workers with ease, but the state provides extensive reemployment support through welfare and training programs. Around 80 percent of the labor force is covered by some form of collective bargaining agreement. Unemployment insurance gives workers 90 percent of their salary for two years, and an elaborate set of government, union, and corporate policies allows almost any employee to attend paid training and pick up new skills. The combination has made the Danish economy uniquely flexible and uniquely equal, and is now warmly embraced by Danish business.

For example, in 2017 the Danish government convened a "Disruption Council" to generate recommendations as to how Denmark should handle the accelerating impact of digital technology on the Danish economy. The council was headed by the prime minister and included not only eight ministers but also another thirty members drawn from across Danish society, including CEOs, social partners, experts, and entrepreneurs. It drew up an extensive set of recommendations in four areas: education and training, new labor market institutions, globalization, and "productive and responsible business." This latter area included the first formal agreement between a digital platform and its employees. The agreement ensured the platform's users have the right to—among other things—pensions and a holiday allowance.

We can take at least three lessons from the Danish experience. The first is a reminder of the ability of inclusive institutions to drive prosperity. Denmark is a tiny country with no significant natural resources, yet its social and political institutions have made it one of the richest countries in the world. The second is the importance of the complementarity between the market and the state. Denmark is simultaneously fiercely committed to the power of the free market and to the role of the government in ensuring economic opportunity and social well-being. Indeed the combination of national health care and extensive government-supported job retraining arguably increases the power and flexibility of the free market because by reducing the risks inherent in leaving a job, it makes it much easier for employees to change positions and to start new firms.

The third lesson is the most critical. The case of Denmark highlights how business can play a central role in framing policy without subverting the democratic process. Business is an important and active voice in the conversation, but it does not seek to control either the process or the end point. Its first priority is the health of the country, not immediate financial returns. And for over a hundred years this commitment has been fundamental to its success.

Mauritius: A Particularly Unlikely Success Story

Mauritius was first settled by the Dutch. They imported the first enslaved people, stripped the island of its ebony trees, and killed the last of the dodoes. The French arrived in 1721. They imported more enslaved people and began the large-scale cultivation of sugar. The British took control of the island in 1814, but they used the island only as a way station on the route to India, and left the French elite in control. Slavery was abolished in 1835, and the local landowners turned instead to the importation of indentured labor. Between 1834 and 1910, more than 450,000 Hindu and Muslim Indian nationals arrived on the island, and by 1911 nearly 70 percent of the island's population of roughly 369,000 were Indo-Mauritians.

They had no political representation. The British governed the island through a legislative council composed of a small number of elected members and another group nominated by the governor. The franchise was limited to wealthy property owners, or about 2 percent of the population, and the first Indo-Mauritians were not elected to the council until 1926.

In 1937 a fall in the price of sugar led to widespread rioting during which four protestors were killed, and the next year a general strike was forcibly broken by the British. Tension between the Franco-Mauritians who owned the sugar plantations and the Indo-Mauritians who worked them remained high, and prospects for further development seemed slim. In 1962 James Meade, a Nobel Prize–winning economist,[85] published an article entitled, "Mauritius: A Case Study in Malthusian Economics," in which he suggested that Mauritius faced "ultimate catastrophe unless effective birth control can be introduced fairly promptly."

The crisis came in August 1967, when the British insisted on an open election as the price for independence. It was bitterly fought. The Mauritian Labor Party (MLP), made up largely of Indo-Mauritians, faced the Parti Mauricien Social Démocrate (PDSD), a loose coalition between the Franco-Mauritian sugar barons and the significant Creole population—largely French-speaking descendants of the enslaved people who had been brought in to work the sugar plantations.[86]

In the event the MLP and its allies took 55 percent of the vote and thirty-nine out of sixty-two seats. The old elite was deeply disappointed. One prominent Franco-Mauritian entrepreneur later recalled, "The evening of the vote of '67, Gaëtan Duval [the leader of the PDSD] cried. Me, I was afraid. I wondered who would be there to help the Mauritians." Five months later, only six weeks before the formal grant of independence, violent riots between Creoles and Muslims in Port Louis, the capital city, left twenty-nine people dead and hundreds injured. British troops had to be brought in to restore control. Nearly six hundred houses were burned, and there were more than two thousand arrests.[87]

Seewoosagur Ramgoolam, the leader of the MLP, had become the first prime minister of a free Mauritius, but the country seemed at risk of collapse. Facing similar circumstances, the new government in Kenya had created a one-party state. Tanzania had nationalized minority-owned businesses. Uganda had driven the minority out of the country. Ramgoolam did something very different. He reached out to the PDSD, suggesting that they together form a government of national unity.

It was a risky decision. Ramgoolam faced a divided country and a weak economy dependent on a business sector that was deeply suspicious of his motives. Many on the Mauritian left—including many leading Hindu intellectuals—favored nationalization of the big plantations. There was no history of collaboration between government and business—and many members of Ramgoolam's own party were avowed Marxists who were deeply leery of any rapprochement with the "capitalists" who dominated the economy. Satcam Boolell, minister of agriculture, one of Ramgoolam's most trusted colleagues, told a reporter that he supported the coalition government, but under one condition:

> I am a socialist, and for me, the class struggle continues: capitalists against workers. . . . However, for me as for others, the overriding interest of the country comes first. That interest today is to eliminate unemployment, hunger, destitution. My condition is that the large sugar employers, the groups of the Mauritius Sugar Producers Association, give a formal guarantee that they will employ the

> unemployed, those who have been laid off since the introduction of the wage councils, and that they will start up other projects. I will support the coalition if it leads to work for everyone.

The sugar barons decided to cooperate. The two sides negotiated for two years before the PDSD agreed to join the MLP in power. It was a true power-sharing agreement, with key individuals from the PDSD taking control of some of the most important government ministries. In essence, Ramgoolam promised to leave the sugar barons in place if they would not only help him diversify the island's economy but also support him in sharing the gains of development widely. The sugar barons agreed to do all they could to support Mauritian development—and to keep wages high.

It is not clear precisely what led the sugar barons to cooperate with the MLP. I've only been able to glean clues to the possible answer. It may be that the island was small enough—and its elite sufficiently closely connected by shared educational experiences and social ties—to permit the Franco-Mauritians to embrace a commitment to the good of Mauritius as a whole. Ramgoolam was a highly educated man—he graduated with a degree in medicine from the University of London—with a long history of engagement with Mauritian politics; some accounts suggest that he was a personal friend of Claude Noel, a prominent Franco-Mauritian and successful entrepreneur who handled the original negotiations between the MLP and the PDSD. Perhaps the Franco-Mauritians were simply unusually farsighted. The Mauritians themselves—using terms very similar to those used by the Danes—talk about dialogue and compromise as being an essential aspect of the "Mauritian way of doing things."

Whatever the cause, the agreement was enormously successful. Leading Franco-Mauritians began to invest aggressively in international tourism. They also spearheaded the development of Export Processing Zones (EPZs)—an idea that had been rejected by the development community as impracticable.[88] Exports from the EPZs grew over 30 percent annually from 1971 to 1975 and played an important role in diversifying the Mauritian economy away from sugar.

Tension between the elites and those who are less well off, and between Francophones and Hindus, continues to this day. But it has always been confined within essentially well-functioning institutions. Elections have been free and fair and have consistently led to transfers in power—including to the Mauritian Marxist party, the "Mouvement Militant Mauricien," in 1982. The judiciary is independent, and there is a lively free press. Even today there are at least nine daily papers published on the island, including the *India Times*, *Chinese Daily News*, the *Independent Daily*, and *Le Defi Quotidien*, *L'Express*, *Le Mauricien*, and *Le Socialiste*.

These institutions—and the close cooperation on which they are based—are widely regarded as having been instrumental in driving Mauritius's unique combination of economic and social strength. Mauritius ranks twenty-fifth in the World Bank's Ease of Doing Business Index and as the eighth "freest economy in the world."[89] Real GDP grew more than 5 percent per year between 1970 and 2009, and in 2018 GDP per capita was $9,697[90]—just behind Poland, Turkey, and Costa Rica. Between 1962 and 2008 the Gini coefficient dropped from 0.50 to 0.38. (In 2013 the United States had a Gini of 0.41. Germany's Gini was 31.4, and Denmark's was 28.5.)[91] Gender equality has improved, and the poverty rate has fallen from 40 to 11 percent. The country recently ranked 11 out of 102 countries in the OECD Social Institutions and Gender Index and 65 in the Human Development Index—ahead of Mexico, Brazil, and China.[92]

What can we learn from Mauritius's experience? As advertised, it turns out that business can help build inclusive institutions outside Europe, even when the local society is not racially homogenous. It's also a story that illustrates the subtle interplay between economic interest and the shared sense of what is right. In Germany, Denmark, and Mauritius, a strong sense of self-preservation led a ruling elite to agree to institutional arrangements they did not favor, and would almost certainly have vigorously resisted if they could. In each case, these arrangements proved to be very successful—and to engender a common sense of shared destiny that led to these ways being increasingly seen as the right way, the only way, the obvious way to behave.

This interplay between self-interest and a shared sense of the right thing is the energy that is propelling so many firms to explore the first four pieces of a reimagined capitalism—shared value, purpose-driven, rewired finance, and self-regulation—and is the reason I believe that they will increasingly support the fifth—the building of inclusive societies. Purpose-driven firms searching for shared value discover new business models that point the way toward making money, while simultaneously reducing pollution and inequality. They build firms authentically committed to doing the right thing, and tell the world—and their employees—that they are committed to making a difference in the world. They then discover that they need government if they are to meet their commitments. Across the world groups of companies passionately committed to making a difference in the world are discovering that shared value is not enough, that self-regulation is unstable, and that investors are not moving fast enough. They are discovering that without the full cooperation of a functional, transparent government that cares about the welfare of its country and its people, many environmental problems cannot be solved, and one can make only minimal headway against reducing inequality. The efforts of purpose driven firms to drive change are the tinder from which the fire of global political reform could spring. Our situation is quite as dire as that facing Germany in 1945 or Denmark in 1895 or Mauritius in 1967.

IN 1971 FUTURE Supreme Court justice Lewis Powell asserted in a widely circulated piece that became known as "the Powell memo" that the American economic system was under broad attack. It was a time when this charge seemed plausible. Government was popular and strong, and the younger generation was actively challenging capitalism. This attack, Powell maintained, required mobilization for political combat: "Business must learn the lesson . . . that political power is necessary; that such power must be assiduously cultivated; and that when necessary, it must be used aggressively and with determination—without embarrassment and without the reluctance which has been so characteristic of American business." Moreover, Powell stressed, the critical ingredient for success would be organization: "Strength lies in organization, in careful long-range planning and implementation, in consistency of action over an indefinite period of years, in the scale of financing available only through joint effort, and in the political power available only through united action and national organizations."

Business leaders answered this call, and in doing so helped to push support for the free market at the expense of governments to such an extent that they helped drive the explosion in inequality that is fueling the populist dragon that stalks the world today. It is time for a new approach—something as organized and as focused on the long term as Powell recommended, but devoted to very different goals.

I talk regularly with CEOs and ex-CEOs. Some of them are Republicans or Conservatives—others are Democrats of various stripes. By and large they have great values, and they all worry deeply about the state of the world. They understand the viability of the system is at risk. But most of them think it's not their job to do anything about it. They are wrong. Rebuilding our institutions is critical to averting long-term disaster and to creating a world in which business can thrive. It's also critical to building a just and sustainable society. This is the time to act. I don't know what this will look like, but one plausible pilot is already underway. It's called Leadership Now, and it is run by Daniella Ballou-Aares.

In the days after the election of 2016, Daniella Ballou-Aares was besieged by dozens of people from her professional and social networks.[93] She had a background in engineering and strategy consulting, a Harvard MBA, and an enviable track record as an entrepreneur, helping to grow Dalberg—a strategic advisory firm—from a seven-person start-up to a strategic advisory firm with twenty-five offices across the globe. (She said later, "We were too young to be starting a strategy consulting firm, but we did it anyway.") But what made her particularly sought-after following the election was that she had spent the previous five years in government, advising the US secretary of state, seeking to transform the United States' approach to foreign aid, and securing an agreement on the Sustainable Development Goals. The experience had been occasionally inspiring, but it had also left her with a profound unease about the state of America's political institutions, and a concern that few businesspeople she knew were paying attention. In her words,

> Within a few months of joining the government I could see that the system was not working. In the stately meeting rooms of the White House and the State Department, we'd debate important policies and ideas, but it was increasingly clear that there was no path to actually doing most of what we were talking about—that most of the ideas could never get through Congress—and that even if they could, they wouldn't get through the bureaucracy, because the system was so antiquated, and so unable to drive change. Plus Congress had almost no incentives to do anything, given the acceleration of gerrymandering and the kind of manipulation that was starting to happen with campaign finance. Congress basically

> only passed one serious piece of legislation—healthcare—during the Obama administration.

She found the election of 2016 deeply sobering, and she started to talk about what could be done with a group of friends from business school. She said,

> When Trump won, there was all of a sudden a strong interest in government, in why it wasn't working and the risks that implied. We felt we had a moment to capture that fear and attention and use it to engage people in actually being part of fixing what wasn't working. We knew that this wasn't just about Trump or Democrats vs. Republicans. People were asking themselves, "Do I just donate money to lots of different organizations? Do I support candidates? Do I go on marches? How can I really have impact?"

The day after the 2017 Women's March on Washington, Daniella and a group of like-minded Harvard Business School classmates organized a one-day conference to hear from both political and nonprofit leaders. They spent the next six months engaging democracy experts, conducting analyses, and testing the idea of a new political venture with their networks. In mid-2017 they founded a membership organization devoted to "identifying the most impactful ways to engage in the long-term project of fixing our political system." Everyone kicked in a little money to get the new organization—"Leadership Now"—off the ground, and as the organization started to gain traction, Daniella stepped down from Dalberg to take on the CEO role full time.

Leadership Now's goal is to catalyze a new commitment to American democracy and to support its members—who are largely businesspeople—in fulfilling that commitment. Members are given opportunities to learn, engage, and invest their time and resources in political reform. The organization hosts speakers, dinner series, and briefings on topics as diverse as developing new political talent and makes available the latest data on political spending. It has developed a "democracy market map," characterizing the money and the players who are influencing democracy and curated a "democracy investment portfolio," a list of organizations that are fighting for reforms that members might want to support. It identifies political candidates who share its values and recommends them to its members. In the 2018 midterms, thirteen of the nineteen candidates it recommended were elected to Congress—more than half of them women, many with backgrounds in business, and all of them committed to political reform. It holds an annual conference during which members meet to debate strategy, listen to politicians and political experts, and simply get to know each other. The group has already seen substantial increases in financial contributions by its members to organizations and candidates dedicated to reforming the system. Everything is geared toward building a connected community that's willing to work together over the long term and that can become effective advocates for political reform. In Daniella's words,

> This is all around responsibility and commitment. We are not just recruiting people to be members of an organization. We're selecting individuals who demonstrate commitment to being part of fixing their democracy—and who recognize it will take real work over a decade or more to do so.

The group now has a hundred and fifty dues-paying members and a presence in six cities—Boston, Houston, New York City, Los Angeles, San Francisco, and Washington, DC. It is explicitly nonpartisan. Members make a serious commitment to engaging in political reform and subscribe to the group's goals—a commitment to defending and renewing democracy by ending gerrymandering, ensuring voter access, and pursuing campaign finance reform—and to its beliefs that "facts and science matter" and that "diversity is an asset," and to the importance of focusing on the long-term health of the nation and the planet.

There are surely many ways in which the private sector could step up to support the institutions of inclusive societies across the world. Indeed my mailbox is full of news about preliminary plans and thoughtful experiments along these lines. My hope is that Leadership Now will soon be only one effort among many.

8

PEBBLES IN AN AVALANCHE OF CHANGE

Finding Your Own Path Toward Changing the World

To be hopeful in bad times is not just foolishly romantic. . . . Human history is a history not only of cruelty, but also of compassion, sacrifice, courage, kindness. What we choose to emphasize in this complex history will determine our lives. If we see only the worst, it destroys our capacity to do something. If we remember those times and places—and there are so many—where people have behaved magnificently, this gives us the energy to act, and at least the possibility of sending this spinning top of a world in a different direction. And if we do act, in however small a way, we don't have to wait for some grand utopian

> future. The future is an infinite succession of presents, and to live now as we think human beings should live, in defiance of all that is bad around us, is itself a marvelous victory.
>
> —HOWARD ZINN, *YOU CAN'T BE NEUTRAL ON A MOVING TRAIN*, 1994

> One small step for a man, one giant leap for mankind.
>
> —NEIL ARMSTRONG

What will a reimagined capitalism look like? It's impossible to know, of course, but at the risk of seeming utopian, let me paint a picture of how the world might look very different twenty years from now.

In a world that has reimagined capitalism, if you're in business, you work for a high-commitment firm that is deeply rooted in shared values, provides great jobs, and takes for granted the idea that while it is essential to be profitable, the firm's primary goal should be to create value, not to make money at any price. Everyone shares a common understanding of the need to balance short-term returns with the public good and the long-term potential of the business. Firms that deny the reality of climate change, treat their employees badly, or actively support corrupt or oppressive political regimes are shunned by their peers and punished by their investors.

Flexible, collaborative agreements across your industry ensure that every organization is held to common standards, so that there are strong incentives for everyone to race to the top. Consumers refuse to buy from firms that cut corners. Prospective employees routinely check out the environmental and social rankings of the firms they are considering joining, and since your firm is at the leading edge of solving several important problems, it has become a magnet for talent. You and your fellow employees have been able to develop new mechanisms through which to express a strong, collective voice not just within your firm but across the entire industry. This voice is welcomed as an important contributor to the long-term health of the society and the free market.

Wherever it can, your firm works closely with government, cooperating in open, public forums to design flexible policies that maximize economic growth while controlling pollution and reinforcing the health of the broader society and of its institutions. Your firm has done its part to support institutional reform, supporting higher taxes, the suppression of corruption, and full democratic access wherever and whenever possible.

There has been a revival of democratic participation: schools everywhere consider "civics" one of their most important subjects, voter participation rates have skyrocketed, and the public conversation is respectful, fact-based, and extraordinarily lively. Governments everywhere control environmental degradation with market-based policies where possible and with direct regulation where it is not, and invest in the public goods that keep societies strong and markets genuinely free and fair. As more and more firms respond to these incentives and focus on transforming their business models to create great jobs, minimize environmental damage, and create the products and services needed to support a sustainable and equitable world, climate change is slowing, inequality is falling, and economic growth continues to be strong.

Once the business community made a commitment to moving to carbon-free energy, progress has been much faster than anyone expected—the OECD countries are on track to decarbonizing their grids by 2050, and the new capacity being built in Africa, China, India, and Brazil is overwhelmingly carbon free. Agricultural practices have been transformed. A strong commitment to ensuring that the cost of these changes should be equitably shared has led to very large investments in retraining and relocation for those who were most affected. These investments have helped to spread prosperity and to blunt the appeal of authoritarian populism.

The shared recognition that the peace and security of the world depends on giving everyone the ability to participate in the free market has led not only to significant investments in education and health but also to a massive expansion in public/private partnerships designed to spur entrepreneurship and new business development in the context of strong social support. Both public and private investment is increasingly focused on the 85 percent of humanity that lives on less than $8 a day,[1] and on the challenging—but exciting and profitable—opportunities inherent in increasing their standard of living without destroying the biosphere.

By now you're probably thinking that I've drunk too much of that purpose-flavored Kool-Aid. But if we decided to reimagine capitalism, we could. Those of us who are doing well under the current system are probably the least well-positioned to see how rapidly change could come. In the early sixties, for example, when a South African psychologist asked a group of students to predict how politics in South Africa would unfold, roughly 65 percent of the black Africans in the group and 80 percent of those of Indian descent predicted the end of apartheid. But only 4 percent of white Afrikaners made the same prediction.[2]

I think it's entirely possible that we will bring everything down upon our heads. But as you may recall from the prologue, I am hopeful. I think it's also entirely possible that we will turn things around. We have the brains, the technology, and the resources to build a just and sustainable world—and in doing so to create enormous economic growth.

The human race has already accomplished far more difficult things. In 1800, 85 percent of humanity lived in extreme poverty. In 2018, only 9 percent did.[3] In 1800 more than 40 percent of all children died before reaching their fifth birthday. Now only one in twenty-six die so young.[4] My father was born in 1935. In his lifetime the world's population more than tripled—from roughly 2.3 billion to about 7.7 billion. But over the same period, world GDP increased fifteen-fold, and GPD per capita has increased from approximately $3,000 to nearly $15,000.[5] That's enough money to permit every man, woman, and child on the planet to meet the core requirements for human happiness: to have enough to eat, decent shelter, and physical security.

We are more peaceful and inclusive than our ancestors would have believed possible. In 1800 slavery was legal almost everywhere, and women did not have the right to vote. Now forced labor is only legal in three countries, and women can vote everywhere there is voting. Almost no one lived in a democracy in 1800. Now more than half of humanity does so; nearly every child receives some form of primary education; and 86 percent of the world's population can read and write. The young are much more likely to believe that climate change is an immediate threat, to support interracial and gay marriage and the rights of women—and much less likely than their parents to support populist leaders.[6] We did not blow ourselves up during the Cold War. We have eradicated smallpox; flown to the moon; and invented the internet, AI, and the cell phone. We have built hearts in petri dishes and reduced the average price of photovoltaic modules a hundred-fold.[7]

The $12 Trillion Opportunity

Most importantly, there's a great business case for saving the world. Meeting the UN's Sustainable Development Goals is a $12 trillion opportunity.[8] Renewable energy is now a more than $1.5 trillion business[9] and in 2017 generated 26.5 percent of global electricity and accounted for 70 percent of all new power generation capacity.[10] Numbers like this make renewables a job creation machine. More than three million Americans now have jobs in the clean energy sector, more than three times the number employed in the fossil fuel industry.[11] Increasing the efficiency with which energy is used could create thousands of new firms and millions of new jobs and cut the world's energy demands by as much as 50 percent.[12]

Switching from beef to "white meat" such as pork or chicken could cut health costs by $1 trillion a year, significantly lower GHG emissions, and greatly reduce the pressure to find new agricultural land.[13] Plant-based food is now a $4.5 billion business[14] and could be an $85 billion industry by 2030.[15] Wheat farmers in the Netherlands, Germany, and the United Kingdom reap more than four times the harvest from the same area of land as farmers in Russia, Spain, and Romania.[16] Yields in much of Africa are even lower. Quadrupling food production is probably an unrealistic goal, but a number of pilot projects suggest that doubling yields—even in the face of climate stress—is eminently possible.[17] About a third of all the food that is produced globally is lost—to pests or spoilage in the supply chain or as consumer waste. Preventing just a quarter of this loss could feed nearly a billion people a year, save nearly a quarter of a trillion dollars, and significantly reduce global GHG emissions.[18]

These are big numbers. On the ground they look like enormous economic opportunity: hundreds of efforts that could create millions of new jobs. All we need to do is reimagine capitalism. All you need to do is help.

Pebbles in an Avalanche of Change

"What can I do?" is the question I am asked most often and certainly the most important one. It's easy to fall into the trap of thinking that only heroes (and heroines!) can change the world. When we tell the story of the civil rights movement, we talk about Martin Luther King and Rosa Parks. When we talk about the New Deal, we talk about Franklin D. Roosevelt. When historians fifty years from now write the history of how we solved global warming, drastically reduced inequality, and remade our institutions, they will focus on a few key events—perhaps in the winter that three superstorms hit the East Coast of the United States, making fixing global warming a completely bipartisan priority, or in the summer that the harvests failed across Africa, sending millions of people north to Europe, making it clear that everyone on the planet had to be given the tools they need to feed themselves. Perhaps they will tell the story of the CEO who led the coalition that helped negotiate a global labor agreement, or of the Chinese and US presidents who sat down together to make a global wealth tax feasible, or of the leaders of the social movement that made it politically impossible not to solve climate change.

But this focus reflects the structure of our minds and the nature of modern communication, not the way in which change actually happens. We use stories to make sense of the noisy, messy, complicated reality of the world, and stories need central characters—single individuals we can identify with and root for.

The real world doesn't actually work that way. Effective leaders ride the wave of change they find bubbling up around them. Martin Luther King did not create the civil rights movement. It grew from decades of work by thousands of African Americans and their allies, each doing the dangerous and difficult work of standing up for change. Rosa Parks was not a lone heroine who simply decided to stay in her seat one evening. She was a deeply committed civil rights worker whose decision that night was taken in close collaboration with a network of experienced female activists. Nelson Mandela did not single-handedly end apartheid in South Africa. He built on fifty years of struggle in which thousands of people participated and hundreds died.

Remember Erik Osmundsen, the CEO who took a corrupt waste collection company and made it a leader in recycling? Whenever he visits my class, he begins by saying that it's not about him. Instead, he insists, it's about the team of people he works with, the people willing to do the actual work—the often dull, day-to-day work—of cleaning up the waste industry. The media tell us that change is dramatic, driven by individuals, and accomplished in minutes. But real change happens one meeting at a time. Remember Michael Leijnse—a relatively low-level employee whose name rarely surfaced in the press—but who, by spearheading sustainable tea at Lipton, showed that it was both possible and profitable, and in doing so, gave his CEO a reason to believe that he could halve Unilever's environmental footprint while increasing its sales.

When Sophia Mendelsohn started at JetBlue, her job was to design a recycling program. But she took the trouble to meet with everyone she could, seeking to understand how focusing on sustainability could help the company as a whole and trying to ensure that everything she did solved a problem for one of her colleagues. Within a few years she was able to spearhead a major shift in how the company measured and managed itself. Greta Thunberg was a fifteen-year-old schoolgirl when she began protesting climate change outside the Swedish Parliament. If it's really a climate emergency, she said, why aren't we doing anything? A year later an estimated 1.6 million students from 125 countries left school to join a global climate strike. I know of one multinational company that completely transformed its sustainability strategy because its employees were finding it too embarrassing to defend the company's actions to their children.

You are vital, and there is lots that you can do. Let me be precise.

Six Steps to Making a Difference[19]

Discover your own purpose. What is it you hold dear? What are you willing to fight for? What do you value above everything else? Whatever you choose to do, make sure that it aligns with the deepest part of who you are. I'm surprised by how often the purpose-driven leaders I meet are deeply rooted in a faith tradition or in a spiritual practice.

Another route to purpose is to reflect on the ways in which the problems of our current age have echoed through your own life. Perhaps there's a place that you loved that you have lost or that has been destroyed. Perhaps you grew up on the wrong side of the tracks, and you saw some of your friends hurt or killed. Perhaps your family has dealt with illness or discrimination or anger. A lot of us are broken, and in the profound brokenness of the world, we see echoes of our own hurts and losses. We become healers to address both our own wounds and those of others.

Some people fight for their children. Some are motivated simply by a burning sense of what's right. If you don't already have a clear sense of what you want to strive for and why, take time to work on yourself—either alone or with others—to learn more. Driving change is hard work. You'll need to be connected to the fire within if you're not to burn out.

Do something now. Decide to fly or drive less, or make the effort to buy only from companies that treat their employees well. Insulate your house and, if you can, put solar panels on the roof or buy your power from a green energy provider. Calculate your carbon footprint, estimate the amount of damage you're doing, and if you can afford it, commit to offsetting that damage. Taking a first step will lead to more. Doing something that's even a little bit outside your comfort zone will change the way you think about yourself. Making even a small sacrifice will help you persuade yourself that you can make some difference and that your voice counts. Something as simple as eating less meat can help you decide to get more active at work—which in turn often opens the door to signing petitions or protesting.[20]

Since we are social primates, your actions will help persuade others to change their own behavior. In one survey, for example, half of the people who responded who knew someone who had given up flying because of climate change said they flew less as a result.[21]

People eating in a café who were told that 30 percent of Americans had recently decided to eat less meat were twice as likely to order a meatless lunch.[22] The odds of someone buying solar panels go up for each home in the neighborhood that already has them.[23]

Find others who share your goals and hang out with them. You cannot save the world alone. I can't even persuade my husband to turn out the lights every time he leaves a room. (He's working on it.) We all need allies—both because there is strength in numbers and because there is no better antidote to despair than working together with others to drive change. It's not a coincidence that people are far more likely to lose weight if they join a support group like Weight Watchers, or far much likely to stay sober if they join Alcoholics Anonymous.[24] Start a book club. Host a series of dinners. Join a nonprofit whose goals you believe in and work you actively support. Every major political and social movement has been fueled by people who were willing to do the hard work of coming together to support each other in demanding change.

Bring your values to work. Start a new company with a different vision. I've been in many meetings in which the threat from a passionate start-up working on a shoestring is the clinching argument that persuades a much larger firm to embrace the need to change. Robin Chase's tiny start-up, Zipcar, transformed how we think about car ownership. Start-ups like First Solar and Bloom Energy convinced thousands of people that it was possible to make money in renewables and energy conservation, helping to start entirely new industries.

You don't have to be the CEO to drive change. If you work at a large organization, you can be a values-driven "intrapreneur"—someone who sees the opportunity for change and builds a team around it. Pick a problem: Changing the light bulbs? Reducing the risk in the supply chain? Improving productivity by reorganizing work and making the company's purpose more salient? Then find some friends, and work on it. Every successful change comes from a demonstration project. Be the demonstration. Sometime soon someone is going to walk into your office and ask you whether it makes sense to clean up the supply chain, to pay people more, or to give everyone the day off to vote. Ask the questions or do the analysis that pushes the company in the right direction. It's nearly always the people on the ground who know what can be done, not the people in the corner office.

If you're a consultant, push your clients to think about the risks and opportunities that the big problems represent. Be the catalyst for change in how they think about the world. If you're an accountant, do the same thing.

Help rewire the capital markets. Work for an impact investor or a family office or a venture capitalist or a private equity firm that understands that there's lots of money to be made in saving the world. Ray Rothrock, an old friend who works at Venrock Ventures, helped raise hundreds of millions of dollars to fund a company called Tri Alpha Energy that is developing a fusion-based technology that could deliver commercially competitive baseload electric power.[25]

Work for an NGO and shame firms into changing—as Greenpeace does, or help them understand how to go about doing it—as organizations like Proforest or Leaders' Quest do.[26] Michael Peck founded 1worker1vote to support worker-owned cooperatives all over the country. Sara Horowitz founded the Freelancers Union, raising $17 million to start an insurance program for her more than four hundred thousand members, and fighting for better pay and conditions. Nigel Topping runs We Mean Business, a coalition of seven international nonprofits that are working together to catalyze business action against climate change.

Work in government. We won't get far without rebuilding trust in government at every level. Smart, capable people who understand that business can be part of the solution, but that externalities need to be properly priced and that the power of business needs to be balanced by the power of the democracy if the whole society is to thrive will be absolutely central to making this happen.

Get political. I know, the idea can be daunting. But it is absolutely essential. Take courage from the examples of others. Remember what Daniella Ballou-Aares has been able to accomplish in just a few years. Some time ago I had tea with Kelsey Wirth, an old friend who is passionate about global warming. We ranted about how slowly politicians were moving to address it and agreed that finding a way to ratchet up public pressure was absolutely critical. Kelsey speculated that mobilizing mothers might be key, since mothers are willing to do almost anything to take care of their children. I left the tea having had a pleasant grumble. But Kelsey—together with a small group of fellow moms—founded Mothers Out Front—a group that now includes more than nineteen thousand mothers and has teams on the ground in nine states. The group engages mothers—deeply and personally—through individual meetings, house parties, and community meetings, and supports them in becoming politically active in effective ways.

In Massachusetts, for example, there are currently more than twenty-three thousand gas leaks.[27] Natural gas is actually methane—a greenhouse gas that traps eighty-six times more heat than CO_2. Of Massachusetts's greenhouse gas emissions, 10 percent is due to methane and—to add insult to injury—the lost gas costs consumers at least $90 million a year. A group of mothers from Mothers Out Front determined to get these leaks fixed. Members of the group have met with activists, city councilors, and state legislators, pushing for legislation to fix the problem. They persuaded a key member of the Boston City Council to schedule a hearing on whether the council should support action—and filled the hearing with committed mothers demanding change. They threatened one of the state's major utilities with a social media "superstorm." By the end of 2016, thirty-seven Massachusetts cities and towns had passed resolutions in favor of new legislation, and the Massachusetts state legislature recently passed an energy bill that includes many of the key provisions Mothers Out Front has been fighting for.

Politicians tell Kelsey that they never hear from the vast majority of their constituents—and that they are surprisingly open to persuasion when twenty highly committed, articulate people come not only to the first hearing for a bill but to the hearing after that, and to the one after that. The women I've met who work with the group tell me that it is one of the best things in their lives. They enjoy getting to know other mothers. They relish the sense of making a difference. Most of all, they value knowing that they are doing something to make sure that their children inherit a sustainable world.

Find a group that is politically active in a way that makes sense to you, and join them. Push for voter registration, or for a climate tax, or a living wage. Working in a community teaches us that organizing is the founding principle of any kind of social change. We need to learn how to take a goal, break it into its component parts, give the right people ownership of each part, and fight until we see a resolution. People will tell you that it's too late, or that it will never work, or that things will never change. But it's never too late. Things can always get worse. A world that warms by six degrees rather than two will be catastrophically worse off. Change is slow until it is fast. The avalanche looks like nothing but a few pebbles moving until the whole hillside goes.

Take care of yourself and remember to find joy. Don't judge your success by whether you save the world. None of us can. There are nearly eight billion wonderful, amazing, occasionally crazy-making human beings on this planet. Each of us can only do what we can do.

Do you know the story about the young woman who saw a beach covered with thousands of stranded starfish and began to throw them, one by one, back into the sea? They say that her friend laughed at her, saying, "What are you doing? Look at this beach: you can't save all these starfish. You can't even begin to make a difference!" The young woman stopped for a moment, thought, and then leaned down to pick up another starfish. "I don't know about that," she replied, "but I know that I'm making a difference to this one."[28]

You don't need to transform the structure of the modern corporation single-handedly to make a difference. If you can make even a small part of a single firm a better place to work, you will change lives.[29]

I know it's hard. I know the temptation to despair. I read bad news for a living, and sometimes it can be hard to get out of bed. But most of the time this work fills me with joy. I'm married to the love of my life, which helps enormously, but I've also had the opportunity to develop a way of thinking about my own role in the world that keeps me going when I'm tempted to simply lie down.

My first husband was a man called John Huchra, who was born in New Jersey on the wrong side of the tracks. His father was a railroad conductor and his mother was a homemaker. Through raw smarts and drivingly hard work, he became an astronomy professor at Harvard, spending as many as two hundred nights a year observing on the world's great telescopes. He was good at what he did. There's a galaxy called "Huchra's lens" in his honor. Together with two collaborators, he drew a map of the nearby universe that revealed a "Great Wall" of galaxies 600 million light-years long and 250 million light-years wide. It was one of the largest cosmic structures ever discovered, and it changed the face of astronomy. Astronomers had always assumed that if they looked out beyond the Milky Way, galaxies would be spread more or less evenly across the universe. But John's map suggested that the galaxies were instead confined to great sheets arcing around enormous voids millions of light-years across. The discovery made the front page of the *New York Times* and helped lay the foundation for the current dark matter–based view of the universe.[30] John became one of the most highly cited astronomers of the twentieth century.

In 1991, when we had our first date, I was entirely ignorant of all this. He was just some guy that I'd been introduced to. Since we were both academics, I asked him how many papers he had published. He hesitated and guessed that it was something over three hundred. Since I had roughly six published papers at the time, I had to suppress a strong urge to leap to my feet and bolt, but we were married a year later, when he was forty-four. He loved the outdoors—particularly hiking and kayaking—and three years after we were married, our son Harry was born. John hadn't thought he would ever marry, let alone have a son, and he loved Harry with a passion. Some fathers feel ambivalence toward their children. As far as I can tell, John never did. Together we watched movies every Friday night. We bribed Harry with small Lego figures to climb the mountains of New Hampshire. Together we made chocolate cookies and cheesecake and laughed and sat around doing nothing at all.

John became the president of the American Astronomical Association, and in 2006 led the delegation to Prague that formally moved the motion to delist Pluto as a planet. Harry was in the room. In 2009 John was a member of a group that went to Rome to meet the Pope—the Catholic hierarchy being keen to demonstrate its support for astronomy ever since the tricky business with Galileo. John had been born Polish Catholic, and he was tickled pink to have the chance to address the Pope. In October 2010 we went to the twenty-fifth reunion of my Harvard Business School class. I can still remember how happy I was that night. I had just moved from MIT to the Harvard Business School and was enjoying it enormously, our son had just entered high school and was thriving, and I was in love. This is it, I remember thinking. This is what we have both worked so hard for, for so many years. I thought we had mastered life.

Five days later, three weeks before my fiftieth birthday. I came home from a business trip to find John lying on the floor. I thought he was playing with the cats. When he wouldn't move, I called 911, screaming at the people who answered to send an ambulance now, right now—so that they could fix him, wake him, something. . . . When I found him at the hospital—after a nightmare drive more surreal than any dream—he was quite dead. I held his hand. We buried him three days later. Harry was only fourteen.

Losing John is one of the hardest things that has ever happened to me. Ordinary life felt like a betrayal. How could one do things like go to the grocery store when John was dead? The warm, tight circle of our family had been blown open. I felt as though I had gone from living in a beautiful house full of family and friends to camping out in a lean-to on an open plain in the midst of driving rain. Great waves of grief swept over me—I cried nearly every day for at least a year. I envied those who still had their partners, still had their families intact.

But I learned. I learned that I had not paid nearly enough attention. A passionate, funny, kind man had shared his life with me, and I had spent far too much time worrying about whether he would take out the garbage. All that laughter and love was gone, and I hadn't treasured every single moment. I learned that people are far more loving and caring than I had given them credit for. People I hardly knew drove across town to give me lasagna at a time when I could barely speak. It felt as if the world had given way beneath me and as if hundreds of hands had reached out to catch me.

I learned that many worse things happen, all the time. One woman I knew—the mother of one of Harry's school friends—stopped me in the parking lot a few weeks after his funeral and expressed her sympathy for my loss. Then she told me that she was leaving her husband because he'd been beating her for more than fifteen years. A colleague told me he had lost his father when he was six. Another mentioned losing a child.

I learned that it is not death that is the tragedy. It is failing to live that is the tragedy. Everybody dies. But not everybody lives. John threw himself into life. He once flew to California to try to persuade a high school class to take more science courses. He once flew to Mexico the week before Christmas to help a graduate student finish her thesis. In a world in which many people hoard, John gave away his data (and his time) to everyone who asked. He loved our son with a fierce passion that stays with Harry today—Harry has often said that he had a father in a way that many of his friends did not. He did world-class science but never made a fuss about who he was or what he'd done. He immersed himself in the beauty of the world. He would go anywhere and do anything, particularly when it meant hauling fifty pounds of gear uphill in the rain. As far as I could tell, he never cared about money or status. He wanted to do great science, support his students—and anyone who needed his help—and to love the natural world and his family. He gave and gave of himself. The day before his funeral I saw him—I still don't know if it was a dream or a waking vision—walking down a road toward distant mountains. He looked over his shoulder toward me and laughed. "Look for me in the trees and the rain," he said, and set off toward the horizon.

When people ask me what keeps me going, I tell them I'm a Buddhist, and that Buddhism comes with both good news and bad news. The good news is that we're not going to die. The bad news is that this is because we don't exist. I believe—and you should feel free to interpret this as a metaphysical belief, although I believe that it is also a physical fact—that we are not "real" in the way we think we are. We are bundles of very small particles temporarily patterned into structures of swirling energy. We think we are separate. We think we exist. But we are songs the universe is singing—glorious songs—but songs that will end. All we can do is try to sing as best we can.

The roots of our current predicament are fear and separateness. We fear we will never have enough. We feel that we are separate and alone. But we are not. I can't tell you that trying to solve the great problems of our time will make you either rich or famous—although it might. I can tell you that you will have wonderful companions for the journey, that you will feel both more hope and more despair than you expect, and that in the end you will die knowing that you have lived life to the full.

Henry David Thoreau once said, "Most men [and women] lead lives of quiet desperation and go to the grave with the song still in them." But you don't have to. Really.

ACKNOWLEDGMENTS

This book has been more than ten years in the making, and I owe thanks to many people. At MIT, John Sterman was the person who first persuaded me that business could change the world; Bob Gibbons forced me to think clearly about what makes organizations tick; and Nelson Repenning taught me that it's all about making choices. At HBS, Karthik Ramanna and Clayton Rose were extraordinary thinking partners in framing up the core ideas that became *Reimagining Capitalism*.

Joe Lassiter, Mike Toffel, Forest Reinhardt, Jennifer Nash, John Macomber, and Dick Vietor kept me grounded in the realities of climate change and business, while Paul Healy and Nien-he Hsieh helped me think through the intersections between leadership and morality, Mike Beer and Russ Eisenhardt showed me that purpose-driven business was a present reality, and Jane Nelson and John Ruggie explained the importance of public-private partnerships and global institutions.

Beyond HBS, David Moss, Richard Locke, and Luigi Zingales continue to be enduring inspirations of how scholars can shape practice. Bruce Kogut hosted me for an immensely valuable visit to Columbia Business School. Marshall Ganz taught me about narrative and organizing, and kept reminding me that reimagining government is just as important as reimagining business. Ioannis

Ioannou helped me think systematically about purpose and financial performance. Sarah Kaplan gave me the courage to say what I really think. Joshua Gans never gave up believing that architectural innovation was important. Rajendra Sisodia, Carol Sanford, Katrin Kaufer, and Otto Scharmer kept reminding me that the heart is just as important as the head.

Mariana Osequera Rodriguez and Tony He provided invaluable research assistance. Tony, you covered more ground than I would have imagined possible. Mariana, you have the patience of a saint and the work ethic of Thomas Edison. Thank you for helping to carry the book over the finish line. Jessica Gover, Kate Isaacs, Carin Knoop, Amram Migdal, Aldo Sesia, Jim Weber, and Hann-Shuin Yew were wonderful partners in putting cases together. Elliott Stoller and Chris Eaglin helped me see the world through the very different eyes of millennials.

My students have shown me that almost everything is possible. My particular thanks to Ryan Allis, Chelsea Banks, Ruzwana Bashir, Lukas Baumgartner, Oriel Carew, Howard Fisher, Diogo Freire, Casey Gerald, Patrick Hidalgo, Aman Kumar, Sam Lazarus, Craig Matthews, Smriti Mishra, Alison Omens, Paulina Ponce de Leon, Robert Poor, Anne Pratt, Prem Ramaswami, Carmichael Roberts, Adam Siegel, Dorjee Sun, Henry Tsai, and Brian Tomlinson.

This book rests on the shoulders of all the businesspeople who have shown me that capitalism can be reimagined. I'm sorry that I cannot name them all here. The protagonists of my cases were endlessly helpful and always inspiring. My thanks—now and always—to Peter Bloom, Karen Colberg, Ralf Carlton, Suzanne McDowell, Mark Bertolini, Stan Bergman, Erik Osmundsen, Reynir Indhal, Michiel Leijnse, Feike Sijbesma, and Hiro Mizuno.

Paul Polman showed me that it was possible to be passionately committed to solving the problems of the world and also be the highly successful, detail-obsessed leader of a multibillion dollar company. Let me watch them try to do the impossible. Doug McMillan and Kathleen McLaughlin gave me hope. Jon Ayers showed me that even a passionate advocate for shareholder value can come to embrace the importance of shared values. Lauren Booker-Allen, Bob Chapman, Catherine Connolly, Sue Garrard, Dick Gochenaur, Diane Propper de Callejon, Kevin Rabinovitch, Jonathan Rose, Arthur Siegel, Carter Williams, Andrew Winston, and Hugh Welsh showed me how passionate commitment could drive change on the ground. I have benefited enormously from my friends in the nonprofit world who share this agenda, including Craig Altemose, Heather Boushey, Mindy Lubber, Lindsay Levin, Michael Peck, Bill Sharpe, Mark Tercek, Nigel Topping, and Judy Samuelson.

My agent, Daniel Stern, took enormous trouble to find the right home for this book. An amazing team helped me produce and position it. My thanks to Mel Blake, Andrew DeSio, Theresa Diederich, Lindsay Fradkoff, Mark Fortier, Jaime Leifer, Dan Masi, Claire Street, and Brynn Warriner. Shazia Amin was a wonderful copyeditor.

Two of the people I would most like to thank are no longer with us. My father, Mungo Henderson, died just under a year before the book was finished. Dean John McArthur, for whom my chair at Harvard is named, passed away a few months later. Both of them were slightly puzzled by this project, but neither could have been more supportive or affectionate. I miss them both immensely and wish they could have lived to see it finished.

My family and friends have been almost infinitely patient with the energy and effort that I poured into this book at their expense. My thanks to Stephanie Connor, who warned me that any reader could put the book down and watch Netflix at any time, and to Sarah Slaughter, Linda Ugelow, Endre Jobbagy, Sarah Robson, Tamlyn Nall, my mother Marina Henderson, my brother Caspar Henderson, and my son Harry Huchra. Steven Holzman and Andrew Schulert read early drafts and gave me hugely helpful comments. Jim Stone's gentle pressure was exactly what I needed.

There are three people without whom this book would not exist. John Mahaney, my editor, who believed in the book from the beginning and spent untold hours helping me wrestle it into shape. George Serafeim, my partner in crime in teaching Reimagining Capitalism for the last five years, who pushed me harder and more productively than anyone I've ever met, and who is a full partner in many of these ideas. George, if anyone can change the world single-handedly, it's you. And Jim Morone, my husband, who showed me that it could be done, encouraged me when it seemed impossible, and throughout everything reminded me to delight in the sheer beauty of the world and the pleasure of being alive.

NOTES

Prologue

1. Gordon Kelly, "Finland and Nokia: An Affair to Remember," *WIRED*, Oct. 4, 2017, www.wired.co.uk/article/finland-and-nokia; "Nokia Smartphone Market Share History," *Statista*, www.statista.com/statistics/263438/market-share-held-by-nokia-smartphones-since-2007/.

Chapter 1: "When the Facts Change, I Change My Mind. What Do You Do, Sir?"

1. WHO (World Health Organization), "Health Benefits Far Outweigh the Costs of Meeting Climate Change Goals," www.who.int/news-room/detail/05-12-2018-health-benefits-far-outweigh-the-costs-of-meeting-climate-change-goals; Intergovernmental Panel on Climate Change (IPCC), *Climate Change 2014: Impacts, Adaptation, and Vulnerability. Part A: Global and Sectoral Aspects. Contribution of Working Group II to the Fifth Assessment Report of the Intergovernmental Panel on Climate Change*, edited by C. B. Field, V. R. Barros, D. J. Dokken, K. J. Mach, M. D. Mastrandrea, T. E. Bilir, M. Chatterjee, K. L. Ebi, Y. O. Estrada, R. C. Genova, B. Girma, E. S. Kissel, A. N. Levy, S. MacCracken, P. R. Mastrandrea, and L. L.White (Cambridge, UK, and New York: Cambridge University Press, 2014).

2. IPCC, *Climate Change 2014*; WWAP (UNESCO World Water Assessment Programme), *The United Nations World Water Development Report 2019: Leaving No One Behind* (Paris: UNESCO, 2019), www.unenvironment.org/news-and-stories/press-release/half-world-face-severe-water-stress-2030-unless-water-use-decoupled.

3. K. K. Rigaud, A. de Sherbinin, B. Jones, J. Bergmann, V. Clement, K. Ober, J. Schewe, S. Adamo, B. McCusker, S. Heuser, and A. Midgley, *Groundswell: Preparing for Internal Climate Migration* (Washington, DC: World Bank, 2018).

4. Brooke Jarvis, "The Insect Apocalypse Is Here," *New York Times*, Nov. 27, 2018, www.nytimes.com/2018/11/27/magazine/insect-apocalypse.html.

5. S. Díaz, J. Settele, E. S. Brondizio, H. T. Ngo, M. Guèze, J. Agard, A. Arneth, et al., eds., "Summary for Policymakers of the Global Assessment Report on Biodiversity and Ecosystem Services of the Intergovernmental Science-Policy Platform on Biodiversity and Ecosystem Services" (Bonn, Germany: IPBES Secretariat, 2019).

6. Hans Rosling, Ola Rosling, and Anna Rosling Rönnlund, *Factfulness: Ten Reasons We're Wrong About the World—and Why Things Are Better Than You Think*, 1st ed. (New York, Flatiron Books, 2018).

7. WHO, "World Bank and WHO: Half the World Lacks Access to Essential Health Services, 100 Million Still Pushed into Extreme Poverty Because of Health Expenses," Dec. 13, 2017, www.who.int/news-room/detail/13-12-2017-world-bank-and-who-half-the-world-lacks-access-to-essential-health-services-100-million-still-pushed-into-extreme-poverty-because-of-health-expenses; Kate Hodal, "Hundreds of Millions of Children in School but Not Learning," *Guardian*, Feb. 2, 2018, www.theguardian.com/global-development/2018/feb/02/hundreds-of-millions-of-children-in-school-but-not-learning-world-bank; United Nations, "Lack of Quality Opportunities Stalling Young People's Quest for Decent Work—UN Report / UN News," Nov. 21, 2017, https://news.un.org/en/story/2017/11/636812-lack-quality-opportunities-stalling young-peoples-quest-decent work-un-report; James Manyika et al., "Jobs Lost, Jobs Gained: Workforce Transitions in a Time of Automation," McKinsey Global Institute (2017).

8. Steven Levitsky and Daniel Ziblatt, *How Democracies Die*, 1st ed. (New York: Crown Publishing, 2018); Yascha Mounk, *The People vs. Democracy: Why Our Freedom Is in Danger and How to Save It* (Cambridge, MA: Harvard University Press, 2018).

9. "GDP per capita (Current US$)," World Bank Data, https://data.worldbank.org/indicator/NY.GDP.MKTP.CD; "Population, Total," World Bank Data, https://data.worldbank.org/indicator/SP.POP.TOTL; "GDP per Capita (Current US$)," World Bank Data, https://data.worldbank.org/indicator/NY.GDP.PCAP.CD.

10. Larry Fink, "A Sense of Purpose," BlackRock, www.blackrock.com/hk/en/insights/larry-fink-ceo-letter.

11. ICBC (Industrial and Commercial Bank of China) is the world's largest bank.

12. It turns out this story is almost certainly a myth. Billy Perrigo, "Did Martin Luther Nail His 95 Theses to the Church Door?" *Time*, Oct. 31, 2017, https://time.com/4997128/martin-luther-95-theses-controversy/.

13. "Business Roundtable Redefines the Purpose of a Corporation to Promote 'An Economy That Serves All Americans,'" *Business Roundtable*, Aug. 19, 2019, www.businessroundtable.org/business-roundtable-redefines-the-purpose-of-a-corporation-to-promote-an-economy-that-serves-all-americans.

14. "Council of Institutional Investors Responds to Business Roundtable Statement on Corporate Purpose," Council of Institutional Investors, Aug. 19, 2019, www.cii.org/aug19_brt_response.

15. Andrew Pollack, "Drug Goes from $13.50 a Tablet to $750, Overnight," *New York Times*, Sept. 20, 2015, www.nytimes.com/2015/09/21/business/a-huge-overnight-increase-in-a-drugs-price-raises-protests.html.

16. Kate Gibson, "Martin Shkreli: I Should've 'Raised Prices Higher,'" CBS News, CBS Interactive, Dec. 4, 2015, www.cbsnews.com/news/martin-shkreli-i-shouldve-raised-prices-higher/.

17. Stephanie Clifford, "Martin Shkreli Sentenced to 7 Years in Prison for Fraud," *New York Times*, Mar. 9, 2018, www.nytimes.com/2018/03/09/business/martin-shkreli-sentenced.html.

18. Gretchen Morgenson, "Defiant, Generic Drug Maker Continues to Raise Prices," *New York Times*, Apr. 14, 2017, www.nytimes.com/2017/04/14/business/lannett-drug-price-hike-bedrosian.html.

19. Joyce Geoffrey et al., "Generic Drug Price Hikes and Out-of-Pocket Spending for Medicare Beneficiaries," *Health Affairs* 37, no. 10 (2018): 1578–1586.

20. Danny Hakim, Roni Caryn Rabin, and William K. Rashbaum, "Lawsuits Lay Bare Sackler Family's Role in Opioid Crisis," *New York Times*, Apr. 1, 2019, www.nytimes.com/2019/04/01/health/sacklers-oxycontin-lawsuits.html.

21. "Big Oil's Real Agenda on Climate Change," Influence Map, 2019, https://influencemap.org/report/How-Big-Oil-Continues-to-Oppose-the-Paris-Agreement-38212275958aa21196dae3b76220bddc.

22. Anne Elizabeth Moore, "Milton Friedman's Pencil," *The New Inquiry*, Apr. 18, 2017, https://thenewinquiry.com/milton-friedmans-pencil/.

23. Sam Costello, "Where Is the IPhone Made? (Hint: Not Just China)," *Lifewire*, Apr. 8, 2019, www.lifewire.com/where-is-the-iphone-made-1999503.

24. For an early articulation of this model, see, e.g., G. Stigler, *The Theory of Price* (London: Macmillan, 1952).

25. Christina D. Romer and Richard H. Pells, "Great Depression," *Encyclopædia Britannica*, Oct. 16, 2019, www.britannica.com/event/Great-Depression; "Unemployment Rate for United States," *FRED*, Aug. 17, 2012, https://fred.stlouisfed.org/series/M0892AUSM156SNBR.

26. There's a lively debate as to whether the focus on shareholder value *caused* this explosion in growth. Other candidates include globalization, major advances in technology, and the spread of the free market more generally.

27. F. Alvaredo, L. Chancel, T. Piketty, E. Saez, and G. Zucman, *World Inequality Report 2018* (Cambridge, MA: The Belknap Press of Harvard University Press, 2018).

28. Alvaredo et al., *World Inequality Report 2018*.

29. Paul R. Epstein, Jonathan J. Buonocore, Kevin Eckerle, Michael Hendryx, Benjamin M. Stout III, Richard Heinberg, Richard W. Clapp, et al., "Full Cost Accounting for the Life Cycle of Coal," *Annals of the New York Academy of Sciences* 1219 (1): 73–98, via Wiley Online Library, accessed February 2017; burning 1 pound of coal emits about 2 pounds of CO_2 depending on the type of coal.

30. WHO, "COP24 Special Report: Health and Climate Change" (2018); Irene C. Dedoussi, et al., "The Co-Pollutant Cost of Carbon Emissions: An Analysis of the US Electric Power Generation Sector," *Environmental Research Letters* 14.9 (2019): 094003; and see, for example, J. Lelieveld, K. Klingmüller, A. Pozzer, R. T. Burnett, A. Haines, and V. Ramanathan, "Effects of Fossil Fuel and Total Anthropogenic Emission Removal on Public Health and Climate," *PNAS* 116, no. 15 (April 9, 2019): 7192–7197.

31. Peabody Energy, *2018 Annual Report,* www.peabodyenergy.com/Peabody/media/MediaLibrary/Investor%20Info/Annual%20Reports/2018-Peabody-Annual-Report-02.pdf?ext=.pdf.

32. "The Carbon Footprint of a Cheeseburger," *SixDegrees*, Sept. 26, 2017, www.sixdegreesnews.org/archives/10261/the-carbon-footprint-of-a-cheeseburger; "GLEAM 2.0—Assessment of Greenhouse Gas Emissions and Mitigation Potential," *Results / Global Livestock Environmental Assessment Model (GLEAM) / Food and Agriculture Organization of the United Nations*, FAO, www.fao.org/gleam/results/en/.

33. *CEMEX Carbon Disclosure Project Annual Report,* 2018.

34. 46m tonnes CO_2e * $80/ton * 1.1 tonnes/ton.

35. *CEMEX Annual Report,* 2018, www.cemex.com/investors/reports/hom#navigate.

36. *Climate Change*, Marks and Spencer, https://corporate.marksandspencer.com/sustainability/business-wide/climate-change.

37. *Key Facts*, Marks and Spencer, https://corporate.marksandspencer.com/investors/key-facts.

38. Hans Rosling et al., *Factfulness*.

39. Alvaredo et al., *World Inequality Report 2018*.

40. Raj Chetty, "Improving Opportunities for Economic Mobility: New Evidence and Policy Lessons," *Bridges* (Fall 2016).

41. Raj Chetty et al., *Mobility Report Cards: The Role of Colleges in Intergenerational Mobility*, NBER Working Paper no. w23618 (Cambridge, MA: National Bureau of Economic Research, 2017).

42. "Disparities in Life Expectancy in Massachusetts Driven by Societal Factors," Harvard T. H. Chan School of Public Health News, Dec. 19, 2018, www.hsph.harvard.edu/news/hsph-in-the-news/life-expectancy-disparities-massachusetts-societal-factors/; https://data.worldbank.org/indicator/sp.dyn.le00.in.

43. "Too Much of a Good Thing," *Economist*, Mar. 26, 2016, www.economist.com/briefing/2016/03/26/too-much-of-a-good-thing.

44. Ben Casselman, "A Start-up Slump Is a Drag on the Economy. Big Business May Be to Blame," *New York Times*, Sept. 20, 2017, www.nytimes.com/2017/09/20/business/economy/startup-business.html?module=inline.

45. Alan B. Krueger, "Reflections on Dwindling Worker Bargaining Power and Monetary Policy," *Luncheon Address at the Jackson Hole Economic Symposium* 24 (2018); Jan De Loecker and Jan Eeckhout, *The Rise of Market Power and the Macroeconomic Implications*, NBER Working Paper no. w23687 (Cambridge, MA: National Bureau of Economic Research, 2017).

46. Martin Gilens and Benjamin I. Page, "Testing Theories of American Politics: Elites, Interest Groups, and Average Citizens," *Perspectives on Politics* 12, no. 3 (2014): 564–581.

47. Jacob Hartmann, "Disney's Fight to Keep Mickey," Chicago Stigler Center Case no. 3 (November 2017).

48. "Lobbying Spending Database—Walt Disney Co, 1998," OpenSecrets.org.

49. In 1997 Disney's net income from "creative content" was $878 million. *Walt Disney Company 1997 Annual Report*, https://ddd.uab.cat/pub/decmed/46860/iaDISNEY

a1997ieng.pdf. Assuming that 50 percent of this income would have been lost starting in 2023 without the passage of the bill, and that with the passage of the bill it will be retained, and discounting those future income streams at 6 percent.

50. Tim Lee, "15 Years Ago, Congress Kept Mickey Mouse out of the Public Domain. Will They Do It Again?" *Washington Post*, Apr. 23, 2019, www.washingtonpost.com/news/the-switch/wp/2013/10/25/15-years-ago-congress-kept-mickey-mouse-out-of-the-public-domain-will-they-do-it-again/.

51. Brief of George A. Akerlof et al. as Amici Curiae in Support of Petitioners, Eric Eldred et al. v. John D. Ashcroft, Attorney General, 537 U.S. 186 (2003).

52. "Fossil Fuel Interests Have Outspent Environmental Advocates 10:1 on Climate Lobbying," *Yale E360*, July 18, 2018, https://e360.yale.edu/digest/fossil-fuel-interests-have-outspent-environmental-advocates-101-on-climate-lobbying; https://influencemap.org/index.html.

53. Hiroko Tabuchi, "The Oil Industry's Covert Campaign to Rewrite American Car Emissions Rules," *New York Times*, Dec. 13, 2018, www.nytimes.com/2018/12/13/climate/cafe-emissions-rollback-oil-industry.html.

54. Following the references above and assuming that the social cost of carbon is $80/ton.

55. Nichola Groom, "Washington State Carbon Tax Poised to Fail after Big Oil Campaign," *Reuters*, Nov. 7, 2018, www.reuters.com/article/us-usa-election-carbon/washington-state-carbon-tax-poised-to-fail-after-big-oil-campaign-idUSKCN1NC1A9.

56. Jonas Hesse, Mozaffar Khan, and Karthik Ramanna, "Political Standards: Corporate Interest, Ideology, and Leadership in the Shaping of Accounting Rules for the Market Economy," *Journal of Accounting & Economics* 64, no. 20 (2015): 2–3.

57. "U.S. and World Population Clock," *Population Clock*, www.census.gov/popclock/; "Gross Domestic Product," *FRED*, Oct. 30, 2019, https://fred.stlouisfed.org/series/GDP.

58. "Gross Domestic Product for Russian Federation," *FRED*, July 1, 2019, https://fred.stlouisfed.org/series/MKTGDPRUA646NWDB; "Russian Federation," World Bank Data, https://data.worldbank.org/country/russian-federation.

59. United Nations, "About the Sustainable Development Goals—United Nations Sustainable Development," www.un.org/sustainabledevelopment/sustainable-development-goals/.

60. Coral Davenport and Kendra Pierre-Louis, "U.S. Climate Report Warns of Damaged Environment and Shrinking Economy," *New York Times*, Nov. 23, 2018, www.nytimes.com/2018/11/23/climate/us-climate-report.html?module=inline.

61. "Migration, Environment and Climate Change (MECC) Division," International Organization for Migration, Feb. 15, 2019, www.iom.int/complex-nexus#estimates.

Chapter 2. Reimagining Capitalism in Practice

1. The bulk of the material that follows is drawn from "Turnaround at Norsk Gjenvinning," by G. Serafeim and S. Gombas, Harvard Business School Case no. 9-116-012 (January 2017).

2. I first heard this from Peter Senge. Thanks, Peter!

3. Rebecca Henderson and Tony L. He, "Shareholder Value Maximization, Fiduciary Duties, and the Business Judgement Rule: What Does the Law Say?" Harvard Business School Background Note 318-097 (January 2018).

4. Global Reporting Initiative, "Sustainability and Reporting Trends in 2025," Global Reporting.org (2015), www.globalreporting.org/resourcelibrary/Sustainability-and-Reporting-Trends-in-2025-2.pdf.

5. Richard Locke, *The Promise and Limits of Private Power: Promoting Labor Standards in a Global Economy* (Cambridge, UK, and New York: Cambridge University Press, 2013).

6. "Trending: Cocoa Giants Embrace Sustainability, but Consumers Remain Key to Lasting Progress," *Sustainable Brands*, Dec. 12, 2017, https://sustainablebrands.com/read/supply-chain/trending-cocoa-giants-embrace-sustainability-but-consumers-remain-key-to-lasting-progress.

7. Rebecca Henderson and Nien-he Hsieh, "Putting the Guiding Principles into Action: Human Rights at Barrick Gold (A)," Harvard Business School Case no. 315-108, March 2015 (Revised December 2017).

8. Yuval N. Harari, *Sapiens, A Brief History of Humankind* (London: Harvill Secker, 2014) is particularly interesting on this point.

9. One report described the dismal conditions of workers in the poultry industry. The average worker had to perform the same task every 20 seconds, processing more than 14,000 chickens per day. The average wage was $11 per hour, while turnover was around 100 percent per year. The employees are not covered by health insurance, despite the fact that injuries are five times more common than in other industries, and are often forced to use diapers, as bathroom breaks are tightly regulated. "Lives on the Line: The High Human Cost of Chicken," Oxfam America, May 23, 2018, www.oxfamamerica.org/livesontheline/.

10. John Miller, *The Glorious Revolution*, 2nd ed. (Harlow, UK: Longman, 1997).

11. *Encyclopedia Britannica*, Massachusetts Bay Colony / Facts, Map, & Significance [online], accessed Oct. 22, 2019, www.britannica.com/place/Massachusetts-Bay-Colony.

Chapter 3. The Business Case for Reimagining Capitalism

1. Brian Eckhouse, "Solar Beats Coal on U.S. Jobs," Bloomberg.com, May 16, 2018, www.bloomberg.com/news/articles/2018-05-16/solar-beats-coal-on-u-s-jobs.

2. Richard Vietor, "Clean Energy for the Future," Harvard Business School (HBS) Technical Note (August 2019).

3. Ian Johnston, "India Just Cancelled 14 Huge Coal-Fired Power Stations as Solar Energy Prices Hit Record Low," *Independent*, May 24, 2017, www.independent.co.uk/environment/india-solar-power-electricity-cancels-coal-fired-power-stations-record-low-a7751916.html.

4. Mark Kane, "Global Sales December & 2018: 2 Million Plug-in Electric Cars Sold," *InsideEVs*, Jan. 31, 2019, https://insideevs.com/news/342547/global-sales-december-2018-2-million-plug-in-electric-cars-sold/.

5. Kate Taylor, "3 Factors Are Driving the Plant-Based 'Meat' Revolution as Analysts Predict Companies Like Beyond Meat and Impossible Foods Could Explode

into a $140 Billion Industry," *Business Insider*, May 24, 2019, www.businessinsider.com/meat-substitutes-impossible-foods-beyond-meat-sales-skyrocket-2019-5. In May 2019, Beyond Meat, which makes a meatless, plant-based burger with something very close to the taste and texture of real beef, had one of the most successful IPOs of the last ten years. On the first day of trading, the stock surged 163 percent, and the company closed that day with a value of $3.83 billion; Bailey Lipschultz and Drew Singer, "Beyond Meat Makes History with the Biggest IPO Pop Since 2008 Crisis," Bloomberg.com, May 2, 2019, www.bloomberg.com/news/articles/2019-05-02/beyond-meat-makes-history-with-biggest-ipo-pop-since-08-crisis.

6. My account of Unilever and its experience in the tea business is drawn from my case: Rebecca Henderson and Frederik Nelleman, "Sustainable Tea at Unilever," HBS Case no. 9-712-438, November 2012.

7. "Tea Consumption by Country," *Statista*, www.statista.com/statistics/940102/global-tea-consumption/.

8. "Tea Market: Forecast Value Worldwide 2017–2024," *Statista*, www.statista.com/statistics/326384/global-tea-beverage-market-size/; Jasan Potts, et al., *The State of Sustainability Initiatives Review 2014: Standards and the Green Economy* (Winnipeg, Canada: International Institute for Sustainable Development, 2014), www.iisd.org/pdf/2014/ssi_2014.pdf; and "Unilever's Tea Beverages Market Share Worldwide 2012–2021," *Statista*, www.statista.com/statistics/254626/unilevers-tea-beverages-market-share-worldwide/.

9. www.walmart.com/ip/Lipton-100-Natural-Tea-Black-Tea-Bags-100-ct/10307788.

10. Jason Clay, *World Agriculture and the Environment* (Washington, DC: Island Press), 102–103.

11. Rachel Arthur, "Tea Production Rises: But FAO Warns of Climate Change Threat," Beveragedaily.com, William Reed Business Media Ltd., May 30, 2018, www.beveragedaily.com/Article/2018/05/30/Tea-production-rises-but-FAO-warns-of-climate-change-threat.

12. Alan Kroeger et al., "Eliminating Deforestation from the Cocoa Supply Chain" (Washington, DC: World Bank, 2017).

13. Columbia Law School Human Rights Institute, *The More Things Change*, Jan. 2014, https://web.law.columbia.edu/sites/default/files/microsites/human-rights-institute/files/tea_report_final_draft-smallpdf.pdf; "Study Report on Tea Plantation Workers-2016-Ilo.org" (2016), www.ilo.org/wcmsp5/groups/public/—asia/—ro-bangkok/—ilo-dhaka/documents/publication/wcms_563692.pdf.

14. Kericho is not, of course, paradise. See Verita Largo and Andrew Wasley, "PG Tips and Lipton Tea Hit by 'Sexual Harassment and Poor Conditions' Claims," *Ecologist*, Nov. 17, 2017, https://theecologist.org/2011/apr/13/pg-tips-and-lipton-tea-hit-sexual-harassment-and-poor-conditions-claims.

15. "Unpacking the Sustainability Landscape," *Nielsen*, Sept. 11, 2018, www.nielsen.com/us/en/insights/reports/2018/unpacking-the-sustainability-landscape.html.

16. "Unpacking the Sustainability Landscape," *Nielsen*.

17. "Global Consumers Seek Companies That Care About Environmental Issues," *Nielsen*, Sept. 11, 2018, www.nielsen.com/us/en/insights/news/2018/global-consumers-seek-companies-that-care-about-environmental-issues.html.

18. In two large-scale field experiments conducted with the apparel manufacturer, Gap, labels with information about a program to reduce water pollution increased sales by 8 percent among female shoppers, although they apparently had no such effect in outlet stores or on male shoppers. J. Hainmueller and M. J. Hiscox, "The Socially Conscious Consumer," *Field Experimental Test of Consumer Support for Fair Labor Standard* (Massachusetts Institute of Technology Political Science Department Working Paper 2012-15, 2012). In one major US grocery store chain, sales of the two most popular bulk coffees sold in the store rose by almost 10 percent when the coffees were labeled as Fair Trade. Jens Hainmueller, Michael J. Hiscox, and Sandra Sequeira, "Consumer Demand for the Fair Trade Label: Evidence from a Field Experiment," *SSRN Electronic Journal* 97, no. 2 (2011): SSRN, and an experiment on eBay suggested that shoppers were willing to pay a 23 percent premium for coffee labeled Fair Trade. M. J. Hiscox, M. Broukhim, and C. Litwin, "Consumer Demand for Fair Trade: New Evidence from a Field Experiment Using eBay Auctions of Fresh Roasted Coffee," *SSRN Electronic Journal*, (2011). See also Maya Singer, "Is There Really Such a Thing as 'Ethical Consumerism'?" *Vogue*, Feb. 5, 2019, www.vogue.com/article/ethical-consumer-rentrayage-batsheva-lidia-may.

19. Tania Braga, Aileen Ionescu-Somers, and Ralf W. Seifert, "Unilever Sustainable Tea Part II: Reaching out to Smallholders in Kenya and Argentina," accessed November 2011, www.idhsustainabletrade.com/idh-publications.

20. "Britain Backs Kenya Tea Farmers," SOS Children's Village, March 14, 2011, www.soschildrensvillages.org.uk/charity-news/archive/2011/03/britain-backs-kenya-tea-farmers.

21. Root Capital was a nonprofit social investment fund that provided financing for rural businesses in developing countries. It invested in a class of capital that sat between microcredit and commercial lending; Tensie Whelan, Rainforest Alliance, interview by author, Cambridge, MA, October 24, 2011.

22. Rebecca M. Henderson and Frederik Nellemann, "Sustainable Tea at Unilever," HBS Case no. 712-438, December 2011 (Revised November 2012).

23. Idem.

24. Which is not to say that Unilever is perfect or that problems haven't come up. In 2011, for example, a Dutch NGO published a report claiming that female employees at Kericho were subject to systematic sexual harassment.

25. "Tea in the United Kingdom," *Euromonitor International*, January 2011, www.euromonitor.com.

26. The best kind, in my view.

27. Using exchange rate of €1 = A$1.31, as of December 2, 2011.

28. "Tea in Italy," February 2011, *Euromonitor International*, accessed November 2011, www.euromonitor.com.

29. "Unilever's Purpose-Led Brands Outperform," Unilever Global Company website, www.unilever.com/news/press-releases/2019/unilevers-purpose-led-brands-outperform.html.

30. Susan Rosegrant, "Wal-Mart's Response to Hurricane Katrina: Striving for a Public-Private Partnership," Kennedy School of Government Case Program

C16-07-1876.0, Case Studies in Public Policy and Management (Cambridge, MA: Kennedy School of Government, 2007).

31. Suzanne Kapner, "Changing of the Guard at Wal-Mart," *CNNMoney*, Cable News Network, Feb. 18, 2009, https://money.cnn.com/2009/02/17/news/companies/kapner_scott.fortune/. My source for much of the story that follows is my case (and the references therein): Rebecca Henderson and James Weber, "Greening Walmart: Progress and Controversy," HBS Case no. 9-316-042, February 2016.

32. Kapner, "Changing of the Guard at Wal-Mart" (2009).

33. "Our History," *Corporate*, https://corporate.walmart.com/our-story/our-history.

34. Business Planning Solutions, "The Economic Impact of Wal-Mart" (Washington, DC, 2005).

35. Henderson and Weber, "Greening Walmart: Progress and Controversy" (Revised February 2017).

36. Joel Makower, "Walmart Sustainability at 10: The Birth of a Notion," *GreenBiz*, November 16, 2015, www.greenbiz.com/article/walmart-sustainability-10-birth-notion.

37. Alison Plyer, "Facts for Features: Katrina Impact" (The Data Center, August 28, 2015), www.datacenterresearch.org/data-resources/katrina/facts-for-impact/.

38. Edward Humes, *Force of Nature: The Unlikely Story of Wal-Mart's Green Revolution* (New York: Harper Business, 2011), 97–99; Michael Barbaro and Justin Gillis, "Wal-Mart at Forefront of Hurricane Relief," *Washington Post*, Sept. 6, 2005, www.washingtonpost.com/archive/business/2005/09/06/wal-mart-at-forefront-of-hurricane-relief/6cc3a4d2-d4f7-4da4-861f-933eee4d288a/.

39. "Former Laggard Wal-Mart Turns into Ethical Leader—Covalence Retail Industry Report 2008," *Covalence SA*, Dec. 11, 2008, www.covalence.ch/index.php/2008/12/11/former-laggard-wal-mart-turns-into-ethical-leader-covalence-retail-industry-report-2008/.

40. G. I. McKinsey, "Pathways to a Low-Carbon Economy. Version 2 of the Global Greenhouse Gas Abatement Cost Curve," *McKinsey & Company, Stockholm* (2009).

41. By Editor, "Commissioning HVAC Systems," *FM Media*, Jan. 22, 2015, www.fmmedia.com.au/sectors/commissioning-hvac-systems/.

42. Robert G. Eccles, George Serafeim, and Tiffany A. Clay, "KKR: Leveraging Sustainability," HBS Case no. 112-032, September 2011 (Revised March 2012).

43. "Global Industrial Energy-Efficiency Services Market Predicted to Exceed USD 10 Billion by 2020: Technavio," *Business Wire*, Dec. 26, 2016; "Europe's Energy Efficiency Services Market to Reach €50 Billion by 2025," Consultancy.eu, Apr. 2, 2019; "A $300 Billion Energy Efficiency Market," CNBC, Mar. 19, 2019, www.cnbc.com/advertorial/2017/09/19/a-300-billion-energy-efficiency-market.html; *Energy Efficiency Market Report 2018* (Paris: International Energy Agency [IEA], 2018), https://webstore.iea.org/download/direct/2369?fileName=Market_Report_Series_Energy_Efficiency_2018.pdf.

44. Adam Tooze, "Why Central Banks Need to Step Up on Global Warming," *Foreign Policy*, Aug. 6, 2019, https://foreignpolicy.com/2019/07/20/why-central-banks-need-to-step-up-on-global-warming/.

45. Tooze (2019).

46. "Florida's Sea Level Is Rising," *Sea Level Rise*, https://sealevelrise.org/states/florida/.

47. Akhilesh Ganti, "What Is a Minsky Moment?" *Investopedia*, July 30, 2019, www.investopedia.com/terms/m/minskymoment.asp; John Cassidy. "The Minsky Moment," *New Yorker*, January 27, 2008, www.newyorker.com/magazine/2008/02/04/the-minsky-moment.

48. Christopher Flavelle, "Bank Regulators Present a Dire Warning of Financial Risks from Climate Change," *New York Times*, Oct. 17, 2019, www.nytimes.com/2019/10/17/climate/federal-reserve-climate-financial-risk.html

49. This story draws heavily on my case "CLP: Powering Asia," George Serafeim, Rebecca Henderson, and Dawn Lau, 9-115-038, February 2015.

50. "The First Mobile Phone Call Was Placed 40 Years Ago Today," Fox News, Dec. 20, 2014, www.foxnews.com/tech/2013/04/03/first-mobile-phone-call-was-placed-40-years-ago-today.html.

51. The levelized cost of electricity is the ratio between lifetime costs to lifetime electricity generation using a discount rate that captures the average cost of capital. International Renewable Energy Agency, "Renewable Power Generation Costs in 2018" (Abu Dhabi: IRENA, 2019).

52. IRENA, "Renewable Power Generation Costs in 2018"; IRENA, "Future of Wind: Deployment, Investment, Technology, Grid Integration and Socio-economic Aspects" (A Global Energy Transformation paper, Abu Dhabi: IRENA).

53. See, for example, "New Energy Outlook 2019: Bloomberg NEF," and McKinsey Energy Insights, Global Energy Perspective, January 2019.

54. "China Pushes Regions to Maximize Renewable Energy Usage," *Reuters*, Aug. 30, 2019, www.reuters.com/article/us-china-renewables/china-pushes-regions-to-maximize-renewable-energy-usage-idUSKCN1VK087.

55. "World Energy Outlook 2017 China: Key Findings," International Energy Agency, www.iea.org/weo/china/.

56. "New Energy Outlook 2019: Bloomberg NEF."

57. AutoGrid, *CLP Holdings Signs Multi-Year Strategic Commercial Agreement with AutoGrid to Deploy New Energy Solutions Across Asia-Pacific Region*, Dec. 12, 2018, www.prnewswire.com/in/news-releases/clp-holdings-signs-multi-year-strategic-commercial-agreement-with-autogrid-to-deploy-new-energy-solutions-across-asia-pacific-region-702571991.html.

58. Nico Pitney, "A Revolutionary Entrepreneur on Happiness, Money, and Raising a Supermodel," *Huffington Post*, Dec. 7, 2017, www.huffingtonpost.com/2015/01/30/robin-chase-life-lessons_n_6566944.html.

59. "Avis Budget Group to Acquire Zipcar for $12.25 Per Share in Cash," *Zipcar*, Jan. 2, 2013, www.zipcar.com/press/releases/avis-budget-group-acquires-zipcar.

60. Jackie Krentzman, "The Force Behind the Nike Empire," *Stanford Magazine*, Jan. 1997, https://alumni.stanford.edu/get/page/magazine/article/?article_id=43087.

61. *Nike Annual Report 1992*, NIKE, https://s1.q4cdn.com/806093406/files/doc_financials/1992/Annual_Report_92.pdf.

62. P/E ratios measured as average annual PE.

63. Edward Yardeni et al., "Stock Market Briefing: S&P 500 Sectors & Industries Forward P/Es," Yardeni.com, Aug. 26, 2019, www.yardeni.com/pub/mktbriefsppe secind.pdf.

64. Something I highly recommend if you're interested in successful entrepreneurship and in what it takes to build a really successful global company. They are all available on Nike's website, and they make for fascinating reading, https://investors.nike .com/investors/news-events-and-reports/default.aspx.

65. Jeffrey Ballinger, "The New Free-Trade Heel," *Harper's Magazine*, Aug. 1992, http://archive.harpers.org/1992/08/pdf/HarpersMagazine-1992-08-0000971.pdf?AWS AccessKeyId=AKIAJXATU3VRJAAA66RA&Expires=1466354923&Signature=Guz AGJL99jmQtdjxkHswI0WLZJA%3D.

66. Mark Clifford, "Spring in Their Step," *Far Eastern Economic Review* 5 (1992): 56–57.

67. Adam Schwarz, "Running a Business," *Far Eastern Economic Review* (June 20, 1991).

68. He mentions that "we answer the overseas questions in a supplement that is included in the annual report meeting," but I have been unable to locate a copy.

69. Richard Locke, *The promise and perils of globalization, the Case of Nike*, MIT Working Paper July 2002, IPC 02-007.

70. John H. Cushman, Jr., "International Business; Nike Pledges to End Child Labor and Apply U.S. Rules Abroad," *New York Times*, May 13, 1998, www.nytimes .com/1998/05/13/business/international-business-nike-pledges-to-end-child-labor -and-apply-us-rules-abroad.html.

71. Amir Ismael, "Making Green: Nike Is the Biggest and Most Sustainable Clothing and Sneaker Brand," *Complex*, June 1, 2018, www.complex.com/sneakers/2015/08 /nike-is-the-most-sustainable-clothing-company.

72. Tim Harford, "Why Big Companies Squander Good Ideas," *Financial Times*, Sept. 6, 2018, www.ft.com/content/3c1ab748-b09b-11e8-8d14-6f049d06439c.

Chapter 4. Deeply Rooted Common Values

1. David Gelles, "He Ran an Empire of Soap and Mayonnaise. Now He Wants to Reinvent Capitalism," *New York Times*, Aug. 29, 2019, www.nytimes.com/2019/08/29 /business/paul-polman-unilever-corner-office.html.

2. The material that follows draws from Rebecca M. Henderson, Russell Eisenstat, and Matthew Preble, HBS Case no. 318-048, February 2018.

3. Knowledge@Wharton, "Aetna CEO Mark Bertolini on Leadership, Yoga, and Fair Wages."

4. James Surowiecki, "A Fair Day's Wage," *New Yorker*, February 2, 2015, www .newyorker.com/magazine/2015/02/09/fair-days-wage.

5. Lisa Rapaport, "U.S. Health Spending Twice Other Countries' with Worse Results," *Reuters*, Mar. 13, 2018, www.reuters.com/article/us-health-spending/u-s-health -spending-twice-other-countries-with-worse-results-idUSKCN1GP2YN.

6. Ajay Tandon et al., "Measuring Overall Health System Performance for 191 Countries" (Geneva: World Health Organization, 2000).

7. Mark Bertolini, *Mission Driven Leadership: My Journey as a Radical Capitalist* (New York: Currency, Penguin Random House, 2019).

8. Jesse Migneault, *Top 5 Largest Health Insurance Payers in the United States*, HealthPayerIntelligence, Apr. 13, 2017, https://healthpayerintelligence.com/news/top-5-largest-health-insurance-payers-in-the-united-states.

9. MarquiMapp, "Aetna CEO Takes Health Care Personally," CNBC, Aug. 3, 2014, www.cnbc.com/2014/08/01/aetna-ceo-takes-health-care-personally.html.

10. Jayne O'Donnell, "Aetna CEO Got Summer's First Merger Agreement, Raised Minimum Wage and More," *USA Today*, Gannett Satellite Information Network, Sept. 8, 2015, www.usatoday.com/story/money/2015/09/07/aetna-ceo-bertolini-yoga-meditation-motorcycles-minimum-wage/29782741/.

11. David Gelles, "Mark Bertolini of Aetna on Yoga, Meditation and Darth Vader," *New York Times*, Sept. 21, 2018, www.nytimes.com/2018/09/21/business/mark-bertolini-aetna-corner-office.html.

12. Meera Viswanathan et al., "Interventions to Improve Adherence to Self-Administered Medications for Chronic Diseases in the United States: A Systematic Review," *Annals of Internal Medicine* 157, no. 11 (2012): 785–795.

13. Idem.

14. Aurel O. Iuga and Maura J. McGuire, "Adherence and Health Care Costs," *Risk Management and Healthcare Policy* 7 (2014): 35.

15. Rebecca M. Henderson, Russell Eisenstat, and Matthew Preble, "Aetna and the Transformation of Health Care," HBS Case no. 318-048, February 2018.

16. Idem.

17. Idem.

18. Rebecca Henderson, "Tackling the Big Problems: Management Science, Innovation and Purpose" (Working paper prepared for Management Science's 65th Anniversary, October 2019).

19. Gelles, "Mark Bertolini of Aetna on Yoga, Meditation and Darth Vader" (2018).

20. Surowiecki, "A Fair Day's Wage" (2015).

21. In 2014, CVS announced that it would stop selling tobacco products, forfeiting roughly $2 billion/year in sales. See Elizabeth Landau, "CVS Stores to Stop Selling Tobacco," CNN, Cable News Network, Feb. 5, 2014, www.cnn.com/2014/02/05/health/cvs-cigarettes/index.html.

22. Jan-Emmanuel De NeveGeorge Ward, "Does Work Make You Happy? Evidence from the World Happiness Report," *Harvard Business Review* (Sept. 20, 2017), https://hbr.org/2017/03/does-work-make-you-happy-evidence-from-the-world-happiness-report.

23. Rebecca Henderson, "Tackling the Big Problems" (October 2019).

24. The description that follows draws on personal communications with KAF executives as well as on the HBS case: Thomas DeLong, James Holian, and Joshua Weiss, "King Arthur Flour," HBS Case no. 9-407-012 (May 2007).

25. www.instagram.com/kingarthurflour/?hl=en.

26. www.facebook.com/GeneralMills/; www.instagram.com/generalmills/; Christian Kreznar, "How King Arthur Flour's Unusual Leadership Structure Is Key to Its Success."

Forbes, Feb. 5, 2019, www.forbes.com/sites/christiankreznar/2019/01/30/how-king-arthur-flours-unusual-leadership-structure-set-it-up-for-success/#48e0e2045c95.

27. "Mission & Impact," King Arthur Flour, www.kingarthurflour.com/about/mission-impact.

28. Alana Semuels, "A New Business Strategy: Treating Employees Well," *Atlantic*, May 7, 2018.

29 "Baker's Hotline," King Arthur Flour, www.kingarthurflour.com/bakers-hotline.

30. www.nationmaster.com/country-info/stats/Economy/GDP-per-capita-in-1950.

31. For references for the account of Toyota and GM that follow, please see Susan Helper and Rebecca Henderson, "Management Practices, Relational Contracts, and the Decline of General Motors," *Journal of Economic Perspectives* 28, no. 1 (2014): 49–72.

32. The workforce management techniques employed by Toyota have been extensively studied by labor economists and specialists in industrial relations. Together they are often called "high-performance work systems." There is no single definition of "high-performance work system," but three overarching elements have been identified in the literature. In general, firms with high-performance work systems (1) implement effective incentive systems, (2) pay a great deal of attention to skills development, and (3) use teams and create widespread opportunities for distributed communication and problem-solving. See, for example, T. A. Kochan, H. C. Katz, and R. B. McKersie, *The Transformation of American Industrial Relations* (New York: Basic Books, 1986); John Paul Macduffie, "Human Resource Bundles and Manufacturing Performance: Organizational Logic and Flexible Production Systems in the World Auto Industry," *Industrial & Labor Relations Review* 48, no. 2 (1995): 197–221; Brian E. Becker et al., "High Performance Work Systems and Firm Performance: A Synthesis of Research and Managerial Implications" (Research in personnel and human resource management, 1998); C. Ichniowski, K. Shaw, and G. Prennushi, "The Effects of Human Resources Management Practices on Productivity: A Study of Steel Finishing Lines," *American Economic Review* 87, no. 3 (1997): 291–314; J. Pfeffer, *The Human Equation* (Boston: Harvard Business School Press, 1998); Eileen Appelbaum et al., *Manufacturing Advantage: Why High-Performance Work Systems Pay Off* (Ithaca, NY: Cornell University Press, 2000); and S. Black and L. Lynch, "How to Compete: The Impact of Workplace Practices and Information Technology on Productivity," *Review of Economics and Statistics* 83, no. 3 (2001): 434–445.

33. Susan Helper and Rebecca Henderson, "Management Practices, Relational Contracts, and the Decline of General Motors," *Journal of Economic Perspectives* 28.1 (2014): 49–72.

34. Benjamin Elisha Sawe, "The World's Biggest Automobile Companies," *World Atlas*, Dec. 13, 2016, www.worldatlas.com/articles/which-are-the-world-s-biggest-automobile-companies.html.

35. Chad Syverson, "What Determines Productivity?" *Journal of Economic Literature* 49, no. 2 (2011): 326–365.

36. Nicholas Bloom and John Van Reenen, "Measuring and Explaining Management Practices Across Firms and Countries," *Quarterly Journal of Economics* 122 (2007): 1351–1408; Bloom and Van Reenen, "Why Do Management Practices Differ Across Firms and Countries?" *Journal of Economic Perspectives* 24, no. 1 (2010): 203–224;

Bloom and Van Reenen, "Human Resource Management and Productivity," in *Handbook of Labor Economics*, vol. 4, ed., Orley Ashenfelter and David Card (Amsterdam: Elsevier and North-Holland, 2011), 1697–1767; Nicholas Bloom et al., "The Impact of Competition on Management Quality: Evidence from Public Hospitals," *The Review of Economic Studies* 82, no. 2 (2015): 457–489; Nicholas Bloom et al. "Does Management Matter? Evidence from India," *Quarterly Journal of Economics* 128, no. 1 (2013): 1–51; Nicholas Bloom, with Erik Brynjolfsson, Lucia Foster, Ron Jarmin, Megha Patnaik, Itay Saporta-Eksten, and John Van Reenen, "What Drives Differences in Management Practices," *American Economic Review* (May 2019).

37. Jim Harter, "Employee Engagement on the Rise in the U.S.," Gallup.com, Aug. 19, 2019, https://news.gallup.com/poll/241649/employee-engagement-rise.aspx.

38. Frederick W. Taylor, "The Principles of Scientific Management" (New York: Harper & Bros., 1911).

39. Charles D. Wrege and Richard M. Hodgetts, "Frederick W. Taylor's 1899 Pig Iron Observations: Examining Fact, Fiction, and Lessons for the New Millennium," *Academy of Management Journal* 43, no. 6 (Dec. 2000): 1283–1291.

40. "NUMMI," *This American Life*, Dec. 14, 2017, www.thisamericanlife.org/403/transcript.

41. One extreme example of this is reported in J. Patrick Wright, *On a Clear Day You Can See General Motors: John Z. DeLorean's Look Inside the Automotive* (New York: Avon, 1979). He reports that at General Motors in the 1970s, it was considered a great honor for a junior executive to be chosen to run the slide presentation at board meetings, but that the executive's career could be ended if he put a slide in the projector incorrectly.

42. Ashley Lutz, "Nordstrom's Employee Handbook Has Only One Rule," *Business Insider*, Oct. 13, 2014, www.businessinsider.com/nordstroms-employee-handbook-2014-10.

43. Robert Spector and Patrick D. McCarthy, *The Nordstrom Way to Customer Service Excellence for Becoming the "Nordstrom" of Your Industry*, 2nd ed. (Hoboken, NJ: John Wiley & Amp Sons, 2012); Christian Conte, "Nordstrom Customer Service Tales Not Just Legend," Bizjournals.com, Sept. 7, 2012, www.bizjournals.com/jacksonville/blog/retail_radar/2012/09/nordstrom-tales-of-legendary-customer.html; Doug Crandall, and Leader to Leader Institute, *Leadership Lessons from West Point*, 1st ed. (San Francisco: Jossey-Bass, 2007).

44. My principle source for the discussion that follows is Christopher Smith, John Child, Michael Rowlinson, and Sir Adrian Cadbury, *Reshaping Work: The Cadbury Experience* (Cambridge, UK: Cambridge University Press, 2009).

45. "Purchase Power of the Pound," Measuring Worth, www.measuringworth.com/calculators/ukcompare/relativevalue.php?use[]=NOMINALEARN&year_early=1861£71&shilling71=&pence71=&amount=8000&year_source=1861&year_result=2018dea.

46. My source for Trist's work and for the discussion of the US developments that followed is Art Kleiner, *The Age of Heretics: A History of the Radical Thinkers Who Reinvented Corporate Management*, 2nd ed. (San Francisco: Jossey-Bass, 2008).

47. My key reference for the history that follows is Kleiner's *The Age of Heretics*.

48. "About i3 Index," Covestro in North America, www.covestro.us/csr-and-sustainability/i3/covestro-i3-index.

49. Personal communication to the author.

50. "Purpose with the Power to Transform Your Organization," *BCG*, www.bcg.com/publications/2017/transformation-behavior-culture-purpose-power-transform-organization.aspx; Alex Edmans, "28 Years of Stock Market Data Shows a Link Between Employee Satisfaction and Long-Term Value," *Harvard Business Review* 24 (Mar. 2016), https://hbr.org/2016/03/28-years-of-stock-market-data-shows-a-link-between-employee-satisfaction-and-long-term-value; Robert G. Eccles, Ioannis Ioannou, and George Serafeim, "The Impact of Corporate Sustainability on Organizational Processes and Performance," *Management Science* 60, no. 11 (November 2014): 2835–2857; Claudine Gartenberg, Andrea Prat, and George Serafeim, "Corporate Purpose and Financial Performance," *Organization Science* 30, no. 1 (January–February 2019): 1–18.

51. "Edelman Trust Barometer Global Report" (2019), https://news.gallup.com/reports/199961/7.aspx.

52. "State of the American Workplace," Gallup.com, May 16, 2019; "Edelman Trust Barometer Global Report" (2019), https://news.gallup.com/reports/199961/7.aspx; "Edelman Trust Barometer, 2019," Edelman, www.edelman.com/sites/g/files/aatuss191/files/2019-02/2019_Edelman_Trust_Barometer_Global_Report.pdf.

53. "The Business Case for Purpose," *Harvard Business Review* (2019), www.ey.com/Publication/vwLUAssets/ey-the-business-case-for-purpose/$FILE/ey-the-business-case-for-purpose.pdf.

Chapter 5. Rewiring Finance

1. Peter J. Drucker, *Managing for the Future: The 1990s and Beyond* (New York: Penguin, 1992).

2. John R. Graham, Campbell R. Harvey, and Shivaram Rajgopal, "The Economic Implications of Corporate Financial Reporting," *Journal of Accounting and Economics* 40, no. 3 (2005): 32–35, fig. 5; John R. Graham, Campbell R. Harvey, and Shivaram Rajgopal, "Value Destruction and Financial Reporting Decisions," *Financial Analysts Journal* 62, no. 6 (Nov. 6, 2006).

3. Board of Governors of the Federal Reserve System 2016, p. 130.

4. Lucian Bebchuk, Alma Cohen, and Scott Hirst, "The Agency Problems of Institutional Investors," *Journal of Economic Perspectives* 31, no. 3 (Summer 2017): 89–112.

5. It was the largest single-day drop in seventeen years of trading. You can see McMillon trying to explain that sometimes it's necessary to take a short-term hit to ensure long-term success to Mad Money's Jim Crammer the evening of the announcement at www.youtube.com/watch?v=4adIq7iJHtc. It's well worth watching.

6. Dominic Barton, "Capitalism for the Long Term," *Harvard Business Review* (March 2011): 85.

7. David Burgstahelr and Ilia Dichev, "Earnings Management to Avoid Earnings Decreases and Losses," *Journal of Accounting and Economics* 24 (1997): 99; John R. Graham, Campbell R. Harvey, and Shiva Rajgopal, "The Economic Implications of

Corporate Financial Reporting," *Journal of Accounting and Economics* 40, nos. 1–3 (2005): 3–73.

8. Katherine Gunny, "The Relation Between Earnings Management Using Real Activities Manipulation and Future Performance: Evidence from Meeting Earnings Benchmarks," 2009, http://ssrn.com/abstract=816025 or http://dx.doi.org/10.2139/ssrn.816025; Paul M. Healy, "The Effect of Bonus Schemes on Accounting Decisions," *Journal of Accounting and Economics* 7 (1985): 85.

9. Joe Nocera, "Wall Street Wants the Best Patents, Not the Best Drugs," Bloomberg.com, Nov. 27, 2018, www.bloomberg.com/opinion/articles/2018-11-27/gilead-s-cures-for-hepatitis-c-were-not-a-great-business-model.

10. Data from Capital IQ.

11. Data from FactSet.

12. www.sec.gov/Article/whatwedo.html#create; the SEC protects investors and helps maintain fair and efficient markets. It oversees the key participants in the securities space, including investors, mutual funds, security exchanges, and brokers and dealers. The SEC has the authority to take civil enforcement actions against businesses and individuals that violate security laws (e.g., inside trading and accounting fraud, among others).

13. Eugene Soltes, *Why They Do It: Inside the Mind of the White-Collar Criminal* (New York: PublicAffairs, 2016).

14. "ESG Sustainable Impact Metrics—MSCI," Msci.Com, 2019, www.msci.com/esg-sustainable-impact-metrics.

15. Alan Taylor, "Bhopal: The World's Worst Industrial Disaster, 30 Years Later," *Atlantic*, December 2014; Adrien Lopez. "20 Years on from Exxon Valdez: What Progress for Corporate Responsibility?" Mar. 29, 2009, www.ethicalcorp.com/communications-reporting/20-years-exxon-valdez-what-progress-corporate-responsibility.

16. Mindy S. Lubber, "30 Years Later, Investors Still Lead the Way on Sustainability," *Ceres*, Mar. 23, 2019, www.ceres.org/news-center/blog/30-years-later-investors-still-lead-way-sustainability. I have been a member of the CERES board since 2017.

17. "GRI at a Glance," Global Reporting Initiative (GRI), www.globalreporting.org/information/news-and-press-center/press-resources/Pages/default.aspx.

18. "Sustainability and Reporting Trends in 2025," Global Reporting.org (2015), www.globalreporting.org/resourcelibrary/Sustainability-and-Reporting-Trends-in-2025-2.pdf.

19. "2018 Global Sustainable Investment Review," Global Sustainable Investment Alliance (2018), www.gsi-alliance.org/wp-content/uploads/2019/03/GSIR_Review2018.3.28.pdf; Renaud Fages et al. "Global Asset Management 2018: The Digital Metamorphosis," www.bcg.com; BCG, www.bcg.com/publications/2018/global-asset-management-2018-digital-metamorphosis.aspx.

20. "2018 Global Sustainable Investment Review"; Fages et al., "Global Asset Management 2018."

21. See, for example, Christophe Revelli and Jean-Laurent Viviani, "Financial Performance of Socially Responsible Investing (SRI): What Have We Learned? A Meta-analysis," *Business Ethics: A European Review* 24, no. 2 (April 2015).

22. Mozaffar Khan, George Serafeim, and Aaron Yoon, "Corporate Sustainability: First Evidence on Materiality," *Accounting Review* 91, no. 6 (November 2016).

23. "Materiality," *Business Literacy Institute Financial Intelligence*, Sept. 23, 2016.

24. Khan et al., "Corporate Sustainability (2016): 1697–1724; Eccles et al., "The Impact of Corporate Sustainability on Organizational Processes and Performance" (2014): 2835–2857.

25. The discussion below draws extensively on Julie Battilana and Michael Norris, "The Sustainability Accounting Standards Board (Abridged)," HBS Case no. 419-058, March 2019.

26. Jean partnered with a number of key thought leaders, including Robert Massie, Bob Eccles, and David Wood. She describes her decision to found SASB as very much a purpose-driven decision. As Jean recalled, "It was scary to stop earning a regular income, but I truly felt a moral responsibility to take this idea forward because it had the potential to have such a major impact in the US and around the world."

27. Under SEC regulations, every publicly traded firm has the duty to alert its investors to any information that is "material," where information is material if there is "a substantial likelihood that the disclosure of the omitted fact would have been viewed by the reasonable investor as having significantly altered the "total mix" of information made available.

28. Khan et al., "Corporate Sustainability."

29. George Serafeim and David Freiberg, "JetBlue: Relevant Sustainability Leadership (A)," HBS Case no. 118-030, October 2018.

30. "Bio," Sophia Mendelsohn, www.sophiamendelsohn.com/bio.

31. JetBlue, "2016 Sustainability Accounting Standards Board Report" (2017), http://responsibilityreport.jetblue.com/2016/JetBlue_SASB_2016.pdf.

32. Serafeim and Freiberg, "JetBlue."

33. The group that handled relationships with investors.

34. Serafeim and Freiberg, "JetBlue."

35. The principle reference for the discussion that follows is Rebecca Henderson, George Serafeim, Josh Lerner, and Naoko Jinjo, "Should a Pension Fund Try to Change the World? Inside GPIF's Embrace of ESG," HBS Case no. 319-067, January 2019 (Revised March 2019).

36. Eric Schleien, "Investing: Buy What You Know," *Guru*, Apr. 9, 2007, www.gurufocus.com/news/5281/investing-buy-what-you-know.

37. "Peter Lynch," *AJCU*, https://web.archive.org/web/20141226131715/www.ajcunet.edu/story?TN=PROJECT-20121206050322; Peter Lynch, "Betting on the Market-Pros," PBS, www.pbs.org/wgbh/pages/frontline/shows/betting/pros/lynch.html. If you had invested $1,000 in the Magellan the day Lynch took over, it would have been worth $28,000 the year he left.

38. Steven Perlberg, "Mutual Fund Legend Peter Lynch Identifies His 'Three C's' of Investing in a Rare Interview," *Business Insider*, Dec. 6, 2013, www.businessinsider.com/peter-lynch-charlie-rose-investing-2013-12.

39. Kenneth R. French, "Presidential Address: The Cost of Active Investing," *Journal of Finance* 63, no. 4 (2008): 1537–1573.

40. Indeed 90 percent of GPIF's Japanese and 86 percent of its foreign equity assets are invested passively.

41. Sean Fleming, "Japan's Workforce Will Be 20% Smaller by 2040," *World Economic Forum*, Feb. 12, 2019, www.weforum.org/agenda/2019/02/japan-s-workforce-will-shrink-20-by-2040/.

42. "The Global Gender Gap Report 2013," *World Economic Forum*, 236, http://www3.weforum.org/docs/WEF_GenderGap_Report_2013.pdf; "The Global Gender Gap Report 2017," *World Economic Forum*, 90, http://www3.weforum.org/docs/WEF_GGGR_2017.pdf.

43. GPIF was allowed to directly invest in bonds and mutual funds; 15 percent of GPIF's fixed-income assets were managed internally.

44. "The Benefits and Risks of Passive Investing," Barclays, www.barclays.co.uk/smart-investor/investments-explained/funds-etfs-and-investment-trusts/the-benefits-and-risks-of-passive-investing/.

45. See GPIF's *2018 Annual Report*.

46. The Nikkei Telecon Database, accessed December 2018.

47. Size is classified based on market capitalization.

48. "2018 Global Sustainable Investment Review," Global Sustainable Investment Alliance (2018), www.gsi-alliance.org/wp-content/uploads/2017/03/GSIR_Review2016.F.pdf.

49. The question of whether family firms outperform publicly traded companies is highly contested, perhaps because it is difficult to collect comprehensive financial information from family-owned firms. Some sources believe that they are more likely to trade off short-term returns for long-term resilience, and that on average they outperform. See, for example, Kate Rodriguez, "Why Family Businesses Outperform Others," *Economist*, https://execed.economist.com/blog/industry-trends/why-family-businesses-outperform-others. Other evidence suggests that they underperform. See, for example, Andrea Prat, "Are Family Firms Damaging Europe's Growth?" *World Economic Forum*, Feb. 12, 2015, www.weforum.org/agenda/2015/02/are-family-firms-damaging-europes-growth/ and the stream of work by Nicholas Bloom and his collaborators referenced above. For more on the governance of family firms and their role in shaping the broader economy see Randall K. Morck, ed., *A History of Corporate Governance Around the World* (Chicago and London: University of Chicago Press, 2005) and Richard F. Doner and Ben Ross Schneiderm "The Middle-Income Trap: More Politics Than Economics," *World Politics* 68, no. 4 (2016): 608–644.

50. Robert S. Harris, Tim Jenkinson, and Steven N. Kaplan, "How Do Private Equity Investments Perform Compared to Public Equity?" *Journal of Investment Management* 14 no. 3 (2016): 1–24; Robert S. Harris, Tim Jenkinson, and Steven N. Kaplan. "Private Equity Performance: What Do We Know?" *Journal of Finance* 69, no. 5 (2014): 1851–1882.

51. The principle source for the discussion that follows is the Harvard Business School case. Rebecca Henderson, Kate Isaacs, and Katrin Kaufer, "Triodos Bank: Conscious Money in Action," HBS Case no. 313-109, March 2013 (Revised June 2013).

52. "About Triodos Bank," Triodos, www.triodos.com/about-us.

53. Triodos Bank, *Annual Report 2018*, www.triodos-im.com/press-releases/2019/triodos-investment-management-in-2018.

54. My understanding is that Triodos decided not to fund the shoe manufacturer.

55. Triodos Bank, *Annual Report 2018*.

56. Lorie Konish, "The Big Wealth Transfer Is Coming. Here's How to Make Sure Younger Generations Are Ready," CNBC, Aug. 12, 2019, www.cnbc.com/2019/08/12/a-big-wealth-transfer-is-coming-how-to-get-younger-generations-ready.html.

57. References for this and much that follows are from my Mondragon case: Rebecca Henderson and Michael Norris, "1Worker1Vote: MONDRAGON in the U.S.," Harvard Business School Teaching Plan 316–176, April 2016.

58. "The Development and Significance of Agricultural Cooperatives in the American Economy," *Indiana Law Journal* 27, no. 3, Article 2 (1952), www.repository.law.indiana.edu/cgi/viewcontent.cgi?article=2352&context=ilj.

59. "The Development and Significance of Agricultural Cooperatives in the American Economy," *Indiana Law Journal*.

60. Leon Stein, *The Triangle Fire*, 1st ed. (Philadelphia: Lippincott, 1962).

61. Steven Deller, Ann Hoyt, Brent Hueth, and Reka Sundaram-Stukel, "Research on the Economic Impact of Cooperatives," University of Wisconsin Center for Cooperatives, June 19, 2009, http://reic.uwcc.wisc.edu/sites/all/REIC_FINAL.pdf.

62. Tony Sekulich, "Top Ten Agribusiness Companies in the World," *Tharawat Magazine* 12 (June 2019), www.tharawat-magazine.com/facts/top-ten-agribusiness-companies/#gs.001anx.

63. "Leading U.S. Commercial Banks by Revenue 2018," *Statista*, www.statista.com/statistics/185488/leading-us-commercial-banks-by-revenue/.

64. Douglas L. Kruse, ed., "Shared Capitalism at Work: Employee Ownership, Profit and Gain Sharing, and Broad-Based Stock Options," National Bureau of Economic Research Conference Report (University of Chicago Press, May 2010).

65. Douglas L. Kruse, Joseph R. Blasi, and Rhokeun Park, "Shared Capitalism in the U.S. Economy," NBER Working paper no. 14225 (Cambridge, MA: NBER, August 2008).

66. Kruse et al., "Shared Capitalism in the U.S. Economy" (2008).

67. Kruse et al., "Shared Capitalism in the U.S. Economy" (2008).

68. Hazel Sheffield, "The Preston Model: UK Takes Lessons in Recovery from Rust Belt Cleveland," *Guardian*, April 11, 2017, www.theguardian.com/cities/2017/apr/11/preston-cleveland-model-lessons-recovery-rust-belt. See also https://thenextsystem.org/learn/stories/infographic-preston-model.

69. Publix, "About Publix," accessed January 2014, www.publix.com/about/CompanyOverview.do.

70. John Lewis Partnership, "About Us," accessed January 2014, www.johnlewispartnership.co.uk/about.html.

71. "About Us," Mondragon Corporation, www.mondragon-corporation.com/en/about-us/.

72. Mondragon Corporation, *Annual Report 2018* (2018), www.mondragon-corporation.com/en/about-us/economic-and-financial-indicators/annual-report/.

73. "Mondragon Corporation, Winner at the Boldness in Business Awards Organized by the Financial Times," MAPA Group, Mar. 27, 2013, www.mapagroup.net/2013/03/mondragon-corporation-winner-at-the-boldness-in-business-awards-organized-by-the-financial-times/.

74. Critics sometimes charge that employee-owners put themselves at risk by holding so much of their wealth in the firm, but the additional compensation appears to more than make up for this effect. See, for example, Peter Kardas, Adria L. Scharf, and Jim Keogh, "Wealth and Income Consequences of ESOPs and Employee Ownership: A Comparative Study from Washington State," *Journal of Employee Ownership Law and Finance* 10, no. 4 (1998).

75. A defined contribution account is the bank account into which employees and employers pay that later covers an employee's pension; ESOP Association data. Analyzed by NCEO, accessed February 2015, www.esopassociation.org/explore/employee-ownership-news/resources-for-reporters.

76. Kruse et al., "Shared Capitalism in the U.S. Economy?" (2008).

77. Colin Mayer, *Prosperity: Better Business Makes the Greater Good* (Oxford, UK: Oxford University Press, 2019); Lynn Stout, *The Shareholder Value Myth: How Putting Shareholders First Harms Investors, Corporations, and the Public*, 1st ed. (San Francisco: Berrett-Koehler Publishers, 2012); Thomas Donaldson and Lee Preston, "The Stakeholder Theory of the Corporation: Concepts, Evidence, and Implications," *Academy of Management Review* 20, no. 1 (1995): 65–91.

78. See, for example, Stout, *The Shareholder Value Myth*; Mayer, *Prosperity*; Leo Strine, *Towards Fair and Sustainable Capitalism* (Research Paper no. 19-39, University of Pennsylvania Law School, Institute for Law and Economics, September 2019), https://ssrn.com/abstract=3461924.

79. See https://benefitcorp.net/. Becoming a benefit corporation is importantly different from becoming a certified B corporation, which requires only that the firm commit to measuring itself through more than financial metrics. See https://bcorporation.net/.

80. "Benefit Corporation Reporting Requirements," Benefit Corporation, https://benefitcorp.net/businesses/benefit-corporation-reporting-requirements.

81. "State by State Status of Legislation," Benefit Corporation, https://benefitcorp.net/policymakers/state-by-state-status.

82. "Benefit Corporations & Certified B Corps," Benefit Corporation, https://benefitcorp.net/businesses/benefit-corporations-and-certified-b-corps.

83. This duty is typically referred to as the director's Revlon duty. See Leo E. Strine, Jr., "Making It Easier for Directors to Do the Right Thing," *Harv. Bus. L. Rev.* 4 (2014): 235.

84. Stout, *The Shareholder Value Myth*.

85. See FAQ, Benefit Corporation, https://benefitcorp.net/faq.

86. "The Rise and Decline of the Japanese Economic 'Miracle,'" *Understanding Australia's Neighbours: An Introduction to East and Southeast Asia* (Cambridge, UK: Cambridge University Press, 2004): 132–148.

87. Cross-shareholding was understood as a demonstration of a desire to develop long-term business relations among corporations. Until the 1990s, life insurance

companies were one of the most significant shareholder groups in Japan. Japanese banks also held significant equity stakes in their debtors.

88. Nishiyama Kengo, "Proxy Voting Trends in 2014 and Outlook in 2015," presentation, Financial Services Agency, Tokyo, July 9, 2013, www.fsa.go.jp/frtc/kenkyu/gijiroku/20140709/01.pdf.

89. "GDP Growth (Annual %)—Japan," World Bank Data, https://data.worldbank.org/indicator/NY.GDP.MKTP.KD.ZG?locations=JP; "United Kingdom," World Bank Data, https://data.worldbank.org/country/united-kingdom.

90. "GDP Growth (Annual %)—Japan," World Bank Data, https://data.worldbank.org/indicator/NY.GDP.MKTP.KD.ZG?locations=JP; "United Kingdom," World Bank Data, https://data.worldbank.org/country/united-kingdom.

91. Jim Rickards, "Japan's in the Middle of Its 3rd 'Lost Decade' and a Recovery Is Nowhere in Sight," *Business Insider*, Mar. 23, 2016, www.businessinsider.com/japans-3rd-lost-decade-recovery-nowhere-in-sight-2016-3.

92. Ito Kunio, "Ito Review of Competitiveness and Incentives for Sustainable Growth: Building Favorable Relationships Between Companies and Investors," Ministry of Economy, Trade and Industry (METI), August 2014, accessed June 2018, www.meti.go.jp/english/press/2014/pdf/0806_04b.pdf, p. 52.

93. Jake Kanter, "Facebook Shareholder Revolt Gets Bloody: Powerless Investors Vote Overwhelmingly to Oust Zuckerberg as Chairman," *Business Insider*, June 4, 2019, www.businessinsider.com/facebook-investors-vote-to-fire-mark-zuckerberg-as-chairman-2019-6.

94. Guest, CIO Central, "Sorry CalPERS, Dual Class Shares Are a Founder's Best Friend," *Forbes*, May 14, 2013, www.forbes.com/sites/ciocentral/2013/05/14/sorry-calpers-dual-class-shares-are-a-founders-best-friend/#aa06d5012d9b.

95. "Supplier Inclusion," https://corporate.walmart.com/suppliers/supplier-inclusion.

96. Adele Peters, "Tesla Has Installed a Truly Huge Amount of Energy Storage," *Fast Company* June 5, 2018.

97. "The Future of Agriculture," *Economist*, May 11, 2016; "Jain Irrigation Saves Water, Increases Efficiency for Smallholder Farmers," *Shared Value Initiative*, www.sharedvalue.org/examples/drip-irrigation-practices-smallholder-farmers.

98. Brad Plumer, "What's Driving the US Solar Boom? A Bit of Creative Financing," *Vox*, Oct. 8, 2014, www.vox.com/2014/10/8/6947939/solar-power-solarcity-loans-leasing-growth-rooftop.

Chapter 6. Between a Rock and a Hard Place

1. Edward Balleisen, "Rights of Way, Red Flags, and Safety Valves: Business Self-Regulation and State-Building in the United States, 1850–1940," *Journal of Sociology* 113 (2007): 297–351.

2. David Batty, "Unilever Targeted in Orangutan Protest," *Guardian*, Apr. 21, 2008, www.theguardian.com/environment/2008/apr/21/wildlife.

3. Rainforest Rescue, "Facts about Palm Oil and Rainforest," accessed February 2015, www.rainforest-rescue.org/topics/palm-oil; Roundtable on Sustainable Palm Oil (RSPO), "Impact Report 2014," accessed February 2015, www.rspo.org/about/impacts.

4. World Wildlife Fund (WWF), "Which Everyday Products Contain Palm Oil?" accessed February 2016, www.worldwildlife.org/pages/which-everyday-products-contain-palm-oil.

5. Mark L. Clifford, *The Greening of Asia* (New York: Columbia University Press, 2015).

6. World Resources Institute (WRI), "With Latest Fires Crisis, Indonesia Surpasses Russia as World's Fourth-Largest Emitter," Oct. 29, 2015, accessed February 2016, www.wri.org/blog/2015/10/latest-fires-crisis-indonesia-surpasses-russia-world%E2%80%99s-fourth-largest-emitter.

7. Raquel Moren-Penaranda et al., "Sustainable Production and Consumption of Palm Oil in Indonesia: What Can Stakeholder Perceptions Offer to the Debate?" *Sustainable Production and Consumption*, 2015, accessed November 2015, http://ac.els-cdn.com/S2352550915000378/1-s2.0-S2352550915000378-main.pdf?_tid=e5ebb192-8e24-11e5-803f. 00000aacb35d&acdnat=1447872663_63b9570718954aefb715def91b9e8331.

8. Ruysschaert Denis and Denis Salles, "Towards Global Voluntary Standard: Questioning the Effectiveness in Attaining Conservation Goals. The Case of the Roundtable on Sustainable Palm Oil (RSPO)," *Ecological Economics* 107 (2014): 438–446.

9. George Monbiot, "Indonesia Is Burning. So Why Is the World Looking Away?" *Guardian*, Oct. 30, 2015, www.theguardian.com/commentisfree/2015/oct/30/indonesia-fires-disaster-21st-century-world-media.

10. Avril Ormsby, "Palm Oil Protests Target Unilever Sites," *Reuters*, Apr. 21, 2008, https://uk.reuters.com/article/uk-britain-unilever/palm-oil-protests-target-unilever-sites-idUKL2153984120080421.

11. Unilever, "Sustainable Palm Oil: Unilever Takes the Lead," 2008, accessed March 2016, www.unilever.com/Images/sustainable-palm-oil-unilever-takes-the-lead-2008_tcm244-424242_en.pdf.

12. "Unilever PLC Common Stock," Nasdaq, www.nasdaq.com/symbol/ul/stock-comparison.

13. Aaron O. Patrick, "Unilever Taps Paul Polman of Nestlé as New CEO," *Wall Street Journal*, Sept. 5, 2008, www.wsj.com/articles/SB122051169481298737.

14. Indrajit Gupta and Samar Srivastava, "A Person of the Year: Paul Polman," *Forbes*, Feb. 28, 2011, www.forbes.com/2011/01/06/forbes-india-person-of-the-year-paul-polman-unilever.html#141d73761053.

15. It's a key ingredient of products like Doritos tortilla chips.

16. They may also raise antitrust issues. Most jurisdictions permit some form of cooperation in the public interest, but all industry self-regulatory efforts pay great attention to ensuring that they are not in breach of the law.

17. Edward J. Balleisen, "Private Cops on the Fraud Beat: The Limits of American Business Self-Regulation, 1895–1932," *Business History Review* 83 (Spring 2009): 119–120, via Academic Search Premier (EBSCOhost), accessed January 2015.

18. The discussion that follows draws heavily on Christine Meisner Rosen, "Businessmen Against Pollution in Late Nineteenth Century Chicago," *Business History Review* 69, no. 3 (1995): 351–397.

19. "The House of Representatives' Selection of the Location for the 1893 World's Fair," US House of Representatives: History, Art & Archives, http://history.house.gov/HistoricalHighlight/Detail/36662?ret=True.

20. "Worlds Columbian Exposition," *Encyclopedia of Chicago*, http://encyclopedia.chicagohistory.org/pages/1386.html.

21. The account that follows draws heavily on Rosen's wonderful article "Businessmen Against Pollution in Late Nineteenth Century Chicago."

22. "Overview," The Consumer Goods Forum, www.theconsumergoodsforum.com/who-we-are/overview/.

23. Ask Nestle CEO to stop buying palm oil from destroyed rainforest, Greenpeace, www.youtube.com/watch?v=1BCA8dQfGi0.

24. Greenpeace, "2010—Nestlé Stops Purchasing Rainforest-Destroying Palm Oil," 2010, accessed March 2016, www.greenpeace.org/international/en/about/history/Victories-timeline/Nestle/.

25. Gavin Neath and Jeff Seabright, Interview by author, June 28, 2015.

26. Greenpeace, "How Palm Oil Companies Are Cooking the Climate," 2007, accessed March 2016, www.greenpeace.org/international/Global/international/planet-2/report/2007/11/palm-oil-cooking-the-climate.pdf.

27. "No Deforestation, No Peat, No Exploitation Policy," Wilmar, accessed February 2016, www.wilmar-international.com/wp-content/uploads/2012/11/No-Deforestation-No-Peat-No-Exploitation-Policy.pdf.

28. "Cargill Marks Anniversary of No-Deforestation Pledge with New Forest Policy," Cargill, September 17, 2015, www.cargill.com/news/releases/2015/NA31891862.jsp.

29. Roundtable on Sustainable Palm Oil (RSPO), "How RSPO Certification Works," accessed February 2016, www.rspo.org/certification/how-rspo-certification-works.

30. Environmental Investigation Agency (EIA), "Who Watches the Watchmen," November 2015, https://eia-international.org/wp-content/uploads/EIA-Who-Watches-the-Watchmen-FINAL.pdf.

31. Rhett Butler, "Despite Moratorium, Indonesia Now Has World's Highest Deforestation Rate," *Mongabay Environmental News*, Nov. 29, 2015, https://news.mongabay.com/2014/06/despite-moratorium-indonesia-now-has-worlds-highest-deforestation-rate/.

32. Mikaela Weisse and Elizabeth Dow Goldman, "The World Lost a Belgium-Sized Area of Primary Rainforests Last Year," World Resources Institute, Apr. 26, 2019, www.wri.org/blog/2019/04/world-lost-belgium-sized-area-primary-rainforests-last-year.

33. "Indonesia, Global Forest Watch," Global Forest Watch.

34. Terry Slavin, "Deadline 2020: 'We Won't End Deforestation Through Certification Schemes,' Brands Admit," http://ethicalcorp.com/deadline-2020-we-wont-end-deforestation-through-certification-schemes-brands-admit.

35. Shofia Saleh et al., "Intensification by Smallholder Farmers Is Key to Achieving Indonesia's Palm Oil Targets," World Resources Institute, Sept. 26, 2018, www.wri.org/blog/2018/04/intensification-smallholder-farmers-key-achieving-indonesia-s-palm-oil-targets; Thontowi Suhada et al., "Smallholder Farmers Are Key to Making the Palm Oil Industry Sustainable," World Resources Institute, Sept. 26, 2018, www.wri.org/blog/2018/03/smallholder-farmers-are-key-making-palm-oil-industry-sustainable.

36. Philip Jacobson, "Golden Agri's Wings Clipped by RSPO in West Kalimantan," *Forest People Programme*, May 8, 2015, www.forestpeoples.org/topics/palm-oil-rspo/news/2015/05/golden-agri-s-wings-clipped-rspo-west-kalimantan; Annisa Rahmawati,

"The Challenges of High Carbon Stock (HCS) Identification Approach to Support No Deforestation Policy of Palm Oil Company in Indonesia: Lesson Learned from Golden-Agri Resources (GAR) Pilot Project," *IMRE Journal* 7 (3), http://tu-freiberg.de/sites/default/files/media/imre-2221/IMREJOURNAL/imre_journal_annisa_final.pdf, accessed March 2016.

37. "Agriculture, Forestry, and Fishing, Value Added (% of GDP)," World Bank Data, https://data.worldbank.org/indicator/NV.AGR.TOTL.ZS?view=chart.

38. "Employment in Agriculture (% of Total Employment) (Modeled ILO Estimate)," World Bank Data, https://data.worldbank.org/indicator/SL.AGR.EMPL.ZS?view=chart.

39. "What Did Indonesia Export in 2017?" *The Atlas of Economic Complexity*, http://atlas.cid.harvard.edu/explore/?country=103&partner=undefined&product=undefined&productClass=HS&startYear=undefined&target=Product&year=2017.

40. "Agriculture, Forestry, and Fishing, Value Added (% of GDP)," World Bank Data, https://data.worldbank.org/indicator/NV.AGR.TOTL.ZS?view=chart; "What Did Malaysia Export in 2017?" The Atlas of Economic Complexity.

41. World Bank, "Program to Accelerate Agrarian Reform (One Map Project)," https://projects.worldbank.org/en/projects-operations/project-detail/P160661?lang=en.

42. Edward Aspinall and Mada Sukmajati, editors, *Electoral Dynamics in Indonesia: Money Politics, Patronage and Clientelism at the Grassroots* (National University of Singapore Press, 2016).

43. Jake Schmidt, "Illegal Logging in Indonesia: Environmental, Economic & Social Costs Outlined in a New Report," *NRDC*, Dec. 15, 2016, www.nrdc.org/experts/jake-schmidt/illegal-logging-indonesia-environmental-economic-social-costs-outlined-new.

44. Greenpeace, "Eating Up the Amazon," 2006, www.greenpeace.org/usa/wp-content/uploads/legacy/Global/usa/report/2010/2/eating-up-the-amazon.pdf.

45. Greenpeace, "10 Years Ago the Amazon Was Being Bulldozed for Soy—Then Everything Changed," 2016, www.greenpeace.org/usa/victories/amazon-rainforest-deforestation-soy-moratorium-success/, accessed June 2018.

46. Greenpeace, "The Amazon Soy Moratorium," accessed May 2018, www.greenpeace.org/archive-international/Global/international/code/2014/amazon/index.html.

47. Greenpeace, "The Amazon Soy Moratorium."

48. Kelli Barrett, "Soy Sheds Its Deforestation Rap," *GreenBiz*, June 6, 2016, www.greenbiz.com/article/soy-sheds-its-deforestation-rap.

49. Matthew McFall, Carolyn Rodehau, and David Wofford, "Oxfam's Behind the Brands Campaign" (Case study, Washington, DC: Population Council, The Evidence Project, 2017).

50. Greenpeace, "Slaughtering the Amazon," 2009, www.greenpeace.org/usa/wp-content/uploads/legacy/Global/usa/planet3/PDFs/slaughtering-the-amazon-part-1.pdf.

51. Beef provides less than 2 percent of the world's calories. "Agriculture at a Crossroads," *Global Agriculture*, www.globalagriculture.org/report-topics/meat-and-animal-feed.html.

52. Hau Lee and Sonali Rammohan, "Beef in Brazil: Shrinking Deforestation While Growing the Industry," Stanford Graduate School of Business Case no. GS88, 2017.

53. Alexei Barrionuevo, "Giants in Cattle Industry Agree to Help Fight Deforestation," *New York Times*, October 6, 2018, www.nytimes.com/2009/10/07/world/americas/07deforest.html.

54. Holly K. Gibbs et al., "Did Ranchers and Slaughterhouses Respond to Zero-Deforestation Agreements in the Amazon?" *Conservation Letters: A Journal of the Society for Conservation Biology*, April 21, 2015, https://onlinelibrary.wiley.com/doi/full/10.1111/conl.12175.

55. Gibbs et al., "Did Ranchers and Slaughterhouses Respond to Zero-Deforestation Agreements in the Amazon?"

56. Tom Phillips, "Bolsonaro Réjects 'Captain Chainsaw' Label as Data Shows Deforestation 'Exploded,'" *Guardian*, Aug. 7, 2019, www.theguardian.com/world/2019/aug/07/bolsonaro-amazon-deforestation-exploded-july-data.

57. Richard M. Locke, *The Promise and Limits of Private Power: Promoting Labor Standards in a Global Economy* (Cambridge, UK: Cambridge University Press, 2013).

58. Matthew Amengual and Laura Chirot, "Reinforcing the State: Transnational and State Labor Regulation in Indonesia," *ILR Review* 69, no. 5 (2016): 1056–1080.

59. Salo V. Coslovsky and Richard Locke, "Parallel Paths to Enforcement: Private Compliance, Public Regulation, and Labor Standards in the Brazilian Sugar Sector," *Politics & Society* 41, no. 4 (2013): 497–526.

60. Joseph V. Rees, *Hostages of Each Other: The Transformation of Nuclear Safety Since Three Mile Island* (University of Chicago Press, 1994).

61. John G. Kemeny, *Report of the President's Commission on the Accident at Three Mile Island: The Need for Change: The Legacy of TMI*, [the Commission]: For Sale by the Supt. of Docs., U.S. G.P.O., 1979.

62. Jennifer F. Brewer, "Revisiting Maine's Lobster Commons: Rescaling Political Subjects," *International Journal of the Commons* 6, no. 2 (2012): 319–343.

63. Both Japan and Germany essentially shut down their nuclear industries after the 2011 tsunami disaster, suggesting that this fear was well-founded. Thomas Feldhoff, "Post-Fukushima Energy Paths: Japan and Germany Compared." *Bulletin of the Atomic Scientists* 70, no. 6 (2014): 87–96.

64. Bruce Barcott, "In Novel Approach to Fisheries, Fishermen Manage the Catch," *Yale E360*, Jan. 2011, https://e360.yale.edu/features/in_novel_approach_to_fisheries_fishermen_manage_the_catch.

65. The details that follow are drawn from Clayton S. Rose and David Lane, "MELF and Business Culture in the Twin Cities (A)," Harvard Business School Case no. 315-078, March 2015.

66. Winters in Minneapolis are famously cold and long.

67. Harvard Business School MELF Case C.

68. Art Rolnick and Rob Grunewald, "Early Childhood Development: Economic Development with a High Public Return," *The Region* 17, no. 4 (2003): 6–12; his reading of the social science literature suggested that efforts to address the problem would plausibly yield a social rate of return of around 16 percent.

69. Charles McGrath, "Pension Funds Dominate Largest Asset Owners," *Pensions & Investments*, Nov. 12, 2018, www.pionline.com/article/20181112/INTERACTIVE/181119971/pension-funds-dominate-largest-asset-owners.

70. "World's Top Asset Managers 2019," *ADV Ratings*, www.advratings.com/top-asset-management-firms.

71. George Serafeim, "Investors as Stewards of the Commons?" Harvard Business School Working Paper no. 18-013, August 2017.

72. Kelly Gilblom, Bloomberg.com, Apr. 11, 2019, www.bloomberg.com/news/features/2019-04-11/climate-group-with-32-trillion-pushes-companies-for-transparency.

73. The "+" stands for the sixty-one additional "focus companies" that were added to the list six months later either because they will be significantly affected by climate change or because they have a particularly important role to play in mitigating it.

74. https://climateaction100.wordpress.com/.

75. "Power Companies Must Accelerate Decarbonisation and Support Ambitious Climate Policy," FT.com, Dec. 20, 2018.

76. "Proposal: Strategy Consistent with the Goals of the Paris Agreement," *Ceres*, https://ceres.my.salesforce.com/sfc/p/#A0000000ZqYY/a/1H000000bxTX/VMk1IZrSUtwbmXzkJ_DVFFsrtiQBpMuOiZMnzu7V7Y8.

Chapter 7. Protecting What Has Made Us Rich and Free

1. Yascha Mounk, *The People vs. Democracy: Why Our Freedom Is in Danger and How to Save It* (Cambridge, MA: Harvard University Press, 2018).

2. "Unlocking the Inclusive Growth Story of the 21st Century: Accelerating Climate Action in Urgent Times" (Washington, DC: New Climate Economy, 2018), https://newclimateeconomy.report/2018/wp-content/uploads/sites/6/2018/09/NCE_2018_FULL-REPORT.pdf.

3. Manuela Andreoni and Christine Hauser, "Fires in Amazon Rain Forest Have Surged This Year," *New York Times*, Aug. 21, 2019, www.nytimes.com/2019/08/21/world/americas/amazon-rainforest.html.

4. Christine Meisner Rosen, "Businessmen Against Pollution in Late Nineteenth Century Chicago," *Business History Review* 69, no. 3 (1995): 351–397.

5. Pablo A. Mitnik and David B. Grusky, "Economic Mobility in the United States" The Pew Charitable Trusts and the Russel Sage Foundation, 2015); John Jerrim and Lindsey Macmillan, "Income Inequality, Intergenerational Mobility, and the Great Gatsby Curve: Is Education the Key?" *Social Forces* 94, no. 2 (December 2015): 505–533; OECD, "A Family Affair: Intergenerational Social Mobility Across OECD Countries," *Economic Policy Reforms* (2010): 166–183.

6. World Bank, *World Development Report 2018: Learning to Realize Education's Promise* (Washington, DC: World Bank, 2018), doi:10.1596/978-1-4648-1096-1. License: Creative Commons Attribution CC BY 3.0 IGO. Children who don't receive an adequate supply of basic nutrients in their first few days of life suffer cognitive and emotional damage that cannot later be repaired. World Bank, *World Development Report 2018*.

7. F. Alvaredo, L. Chancel, T. Piketty, E. Saez, and G. Zucman, *World Inequality Report 2018* (Cambridge, MA: The Belknap Press of Harvard University Press, 2018).

8. "Total Factor Productivity at Constant National Prices for United States," *FRED*, June 11, 2019, https://fred.stlouisfed.org/series/RTFPNAUSA632NRUG.

9. Lawrence Mishel and Jessica Schieder, "CEO Compensation Surged in 2017," *Economic Policy Institute* 16 (2018).

10. Lyndsey Layton, "Majority of U.S. Public School Students Are in Poverty," *Washington Post*, Jan. 16, 2015, www.washingtonpost.com/local/education/majority-of-us-public-school-students-are-in-poverty/2015/01/15/df7171d0-9ce9-11e4-a7ee-526210d665b4_story.html.

11. Bryce Covert, "Walmart's Wage Increase Is Hurting Its Stock Price and That's OK," *Nation*, Oct. 23, 2015, www.thenation.com/article/walmarts-wage-increase-is-hurting-its-stock-price-and-thats-ok/.

12. Walmart, *2018 Annual Report*, https://s2.q4cdn.com/056532643/files/doc_financials/2018/annual/WMT-2018_Annual-Report.pdf.

13. "Inaugural Addresses of the Presidents of the United States: Ronald Reagan," Avalon Project—Documents in Law, History and Diplomacy, https://avalon.law.yale.edu/20th_century/reagan1.asp.

14. A. Winston, "Where the GOP's Tax Extremism Comes From," [online] 2017, accessed Oct. 18, 2019, https://medium.com/@AndrewWinston/where-the-gops-tax-extremism-comes-from-90eb10e38b1c.

15. "Edelman Trust Barometer Global Report" (2019), https://news.gallup.com/reports/199961/7.aspx.

16. Philip Mirowski and Deiter Piehwe, *The Road from Mont Pelerin: The Making of the Neoliberal Thought Collective* (Cambridge, MA: Harvard University Press, 2009).

17. Theda Skocpol and Alexander Hertel-Fernandez, "The Koch Network and Republican Party Extremism," *Perspectives on Politics* 14, no. 3 (2016): 681–699.

18. Daron Acemoglu, Simon Johnson, and James A. Robinson, "The Colonial Origins of Comparative Development: An Empirical Investigation," *American Economic Review* 91, no. 5 (2001): 1369–1401.

19. City-states in ancient Greece, such as Athens, were notable and rare exceptions.

20. Samuel Edward Finer, *The History of Government from the Earliest Times: Ancient Monarchies and Empires*, vol. 1 (Oxford, UK: Oxford University Press, 1997).

21. Brian M. Downing, "Medieval Origins of Constitutional Government in the West," *Theory and Society* 18, no. 2 (1989): 213–247; Daron Acemoglu and James A. Robinson, *Economic Origins of Dictatorship and Democracy* (Cambridge, UK: Cambridge University Press, 2005).

22. Diego Puga and Daniel Trefler, "International Trade and Institutional Change: Medieval Venice's Response to Globalization," *Quarterly Journal of Economics* 129, no. 2 (2014): 753–821.

23. Barrington Moore, *Social Origins of Dictatorship and Democracy: Lord and Peasant in the Making of the Modern World* (Boston: Beacon Press, 1993).

24. Marina Mazzucato, *The Entrepreneurial State: Debunking Public vs. Private Myths* (London: Anthem Press, 2013).

25. Jeffrey Masters, "The Skeptics vs. the Ozone Hole," *Weather Underground*, 10.226.246.28 (1974), www.wunderground.com/resources/climate/ozone_skeptics.asp.

26. *Chemical Week* (New York: McGraw-Hill, July 16, 1975), print.

27. J. P. Glas, "Protecting the Ozone Layer: A Perspective from Industry," in *Technology and Environment*, ed. J. H. Ausubel and H. E. Sladovich (National Academy Press: Washington, DC, 1989), www.wunderground.com/resources/climate/ozone_skeptics.asp.

28. CFCs are particularly potent greenhouse gases.

29. R. Schmalensee and R. N. Stavins, "The SO2 Allowance Trading System: The Ironic History of a Grand Policy Experiment, *Journal of Economic Perspectives* 27. no. 1 (2013): 103–122.

30. Sanjeev Gupta, Hamid Davoodi, and Rosa Alonso-Terme, "Does Corruption Affect Income Inequality and Poverty?" *Economics of Governance* 3, no. 1 (2002): 23–45.

31. "Views on Homosexuality, Gender and Religion," Pew Research Center for the People and the Press, Sept. 18, 2018, www.people-press.org/2017/10/05/5-homosexuality-gender-and-religion/.

32. "Views on Homosexuality, Gender and Religion." Pew Research Center for the People and the Press, www.people-press.org/2017/10/05/5-homosexuality-gender-and-religion; "Global Attitudes Toward Transgender People," *Ipsos*, www.ipsos.com/en-us/news-polls/global-attitudes-toward-transgender-people.

33. "Business Success and Growth Through LGBT–Inclusive Culture," US Chamber Foundation, Apr. 9, 2019, www.uschamberfoundation.org/sites/default/files/Business-Success-Growth-LGBT-Inclusivc-Culture-FINAL-WEB.pdf.

34. "Frequently Asked Questions about Domestic Partner Benefits," Human Rights Campaign, www.hrc.org/resources/frequently-asked-questions-about-domestic-partner-benefits.

35. "Corporate Equality Index 2019," Human Rights Campaign Foundation, Mar. 28, 2019, https://assets2.hrc.org/files/assets/resources/CEI-2019-FullReport.pdf?_ga=2.70189529.856883140.1563932191-499015526.1563932191.

36. "Corporate Equality Index 2019," Human Rights Campaign Foundation.

37. "LGBT People in the United States Not Protected by State Nondiscrimination Statutes" (Los Angeles: The Williams Institute, UCLA, March 2019); "Hate Crimes," FBI, May 3, 2016, www.fbi.gov/investigate/civil-rights/hate-crimes.

38. Kent Bernhard Jr., "Salesforce CEO Marc Benioff Fights Back Against Indiana 'Religious Freedom' Law," *Business Journals*, Mar. 26, 2015, www.bizjournals.com/bizjournals/news/2015/03/26/benioff-salesforce-fights-indiana-religious-law.html.

39. Jim Gardner, "Other Tech Giants Join Salesforce CEO in Slamming New Indiana Law," Bizjournals.com, www.bizjournals.com/sanfrancisco/blog/2015/03/indiana-gays-discrimination-salesforce-apple-yelp.html.

40. The official summary of the amendment suggests that it "indicates that the law related to adjudicating a claim or defense that a state or local law ordinance or other action substantially burdens the exercise of religion of a person: (1) does not authorize a provider to refuse to offer or provide services, facilities, use of public accommodations, goods, employment, or housing to any member or members of the general public; (2) does not establish a defense to it to a civil action or criminal prosecution for refusal by a provider to offer or provide services, facilities, use of public accommodations, goods, employment or housing to any member or members of the general public; and (3) does

not negate any rights available under the Constitution of the State of Indiana," www .documentcloud.org/documents/1699997-read-the-updated-indiana-religious -freedom.html#document/p1.

41. David Gelles, "The C.E.O. Who Stood Up to President Trump: Ken Frazier Speaks Out," *New York Times*, Feb. 19, 2018, www.nytimes.com/2018/02/19/business /merck-ceo-ken-frazier-trump.html.

42. Matthew E. Kahn et al., "Long-term Macroeconomic Effects of Climate Change: A Cross-Country Analysis," NBER Working Paper no. w26167 (Cambridge, MA: National Bureau of Economic Research, 2019).

43. IPCC, "Summary for Policymakers," in *Climate Change 2014: Mitigation of Climate Change. Contribution of Working Group III to the Fifth Assessment Report of the IPCC*, edited by O. Edenhofer, R. Pichs-Madruga, Y. Sokona, E. Farahani, S. Kadner, K. Seyboth, A. Adler, I. Baum, S. Brunner, P. Eickemeier, B. Kriemann, J. Savolainen, S. Schlömer, C. von Stechow, T. Zwickel, and J. C. Minx (Cambridge, UK, and New York: Cambridge University Press, 2014).

44. Dimitri Zenghelis, "How Much Will It Cost to Cut Global Greenhouse Gas Emissions?" The London School of Economics and Political Science Grantham Research Institute on Climate Change and the Environment website, October 27, 2014, eprints .lse.ac.uk/69605; accessed April 2016; Marshall Burke, Solomon M. Hsiang, and Edward Miguel, "Global Non-Linear Effect of Temperature on Economic Production," *Nature* 527 (November 12, 2015): 235–239, www.nature.com/nature/journal/v527/n7577/full /nature15725.html, accessed June 2016.

45. "What Climate Change Means for Utah," EPA, Aug. 2016, https://19january2017 snapshot.epa.gov/sites/production/files/2016-09/documents/climate-change-ut.pdf.

46. "The Utah Way to Achieving 100 Percent Clean Energy," *Sierra Club*, July 1, 2019, www.sierraclub.org/sierra/2019-4-july-august/feature/utah-way-achieving-100-percent -clean-energy.

47. Iulia Gheorghiu, "PacifiCorp Shows 60% of Its Coal Units Are Uneconomic," *Utility Dive*, Dec. 5, 2018, www.utilitydive.com/news/pacificorp-shows-60-of-its -coal-units-are-uneconomic/543566/.

48. "The Utah Way to Achieving 100 Percent Clean Energy," *Sierra Club*.

49. https://thesolutionsproject.org.

50. Helen Clarkson, "One Year on: U.S. Business Is Still Committed to the Paris Agreement," *GreenBiz*, June 1, 2018, www.greenbiz.com/article/one-year-us-business -still-committed-paris-agreement.

51. Michael, "Trump Will Withdraw U.S. from Paris Climate Agreement," *New York Times*, June 1, 2017, www.nytimes.com/2017/06/01/climate/trump-paris-climate -agreement.html.

52. Andrew Winston, "U.S. Business Leaders Want to Stay in the Paris Climate Accord," *Harvard Business Review* (Feb. 27, 2018), https://hbr.org/2017/05/u-s-business -leaders-want-to-stay-in-the-paris-climate-accord.

53. www.wearestillin.com/.

54. Adam Bonica and Michael McFaul, "Opinion / Want Americans to Vote? Give Them the Day off," *Washington Post*, Oct. 11, 2018, www.washingtonpost.com/opinions

/want-americans-to-vote-give-them-the-day-off/2018/10/10/5bde4b1a-ccae-11e8-920f-dd52e1ae4570_story.html?utm_term=.1bec742b2247.

55. opensecrets.org.

56. opensecrets.org.

57. Marianne Bertrand, Matilde Bombardini, Raymond Fisman, and Francesco Trebbi,"Tax-Exempt Lobbying: Corporate Philanthropy as a Tool for Political Influence," NBER Working Paper no. 24451 (Cambridge, MA: NBER, 2018).

58. See, e.g., Nicholas Confessore and Megan Thee-Brenan, "Poll Shows Americans Favor an Overhaul of Campaign Financing," *New York Times*, June 2, 2015, www.nytimes.com/2015/06/03/us/politics/poll-shows-americans-favor-overhaul-of-campaign-financing.html.

59. www.maketimetovote.org/.

60. Abigail J. Hess, "A Record 44% of US Employers Will Give Their Workers Paid Time off to Vote This Year," CNBC, Oct. 31, 2018, www.cnbc.com/2018/10/31/just-44percent-of-us-employers-give-their-workers-paid-time-off-to-vote.html.

61. Tina Nguyen, "Reid Hoffman's Hundred-Million-Dollar Plan to Growth-Hack Democracy," *Vanity Fair*, July 15, 2019, www.vanityfair.com/news/2019/04/linkedin-founder-reid-hoffman-spends-millions-to-grow-democracy.

62. James Rickards, "Rickards: Warren Buffett and Hugo Stinnes," *Darien Times*, January 5, 2015, www.darientimes.com/38651/rickards-warren-buffett-and-hugo-stinnes/.

63. Sanjeev Gupta, Hamid Davoodi, and Rosa Alonso-Terme, "Does Corruption Affect Income Inequality and Poverty?" *Economics of Governance* 3, no. 1 (2002): 23–45.

64. Gerald D. Feldman, "The Social and Economic Policies of German Big Business, 1918–1929," *American Historical Review* 75, no. 1 (1969): 47–55.

65. David R. Henderson, "German Economic Miracle," *Library of Economics and Liberty*, www.econlib.org/library/Enc/GermanEconomicMiracle.html; Wolfgang F. Stolper and Karl W. Roskamp, "Planning a Free Economy: Germany 1945–1960." *Zeitschrift fur die gesamte Staatswissenschaft* 135, no. 3 (1979): 374–404.

66. Michael R. Hayse, *Recasting West German Elites: Higher Civil Servants, Business Leaders, and Physicians in Hesse Between Nazism and Democracy, 1945–1955*, vol. 11 (New York: Berghahn Books, 2003): 119–120.

67. Kathleen Thelen, *How Institutions Evolve: The Political Economy of Skills in Germany, Britain, the United States and Japan* (Cambridge, UK: Cambridge University Press, 2012).

68. This ranking excludes the rich city-states, "Country Comparison: GDP—PER CAPITA (PPP)," Central Intelligence Agency, www.cia.gov/library/publications/the-world-factbook/rankorder/2004rank.html.

69. World Bank, *World Trade Indicators*, 2017, https://wits.worldbank.org/CountryProfile/en/Country/WLD/Year/2017; "U.S. Exports, as a Percentage of GDP," *Statista*, www.statista.com/statistics/258779/us-exports-as-a-percentage-of-gdp/.

70. Klaus Schwab, "The Global Competitiveness Report 2018," *World Economic Forum*, 2018.

71. "Infrastructure," Germany Trade and Invest GmbH (GTAI), www.gtai.de/GTAI/Navigation/EN/Invest/Business-location-germany/Business-climate/infrastructure.html.

72. Hermann Simon, "Why Germany Still Has So Many Middle-Class Manufacturing Jobs," *Harvard Business Review* (July 13, 2017), https://hbr.org/2017/05/why-germany-still-has-so-many-middle-class-manufacturing-jobs?referral=03759&cm_vc=rr_item_page.bottom.

73. Tamar Jacoby, "Why Germany Is So Much Better at Training Its Workers," *Atlantic*, Oct. 20, 2014, www.theatlantic.com/business/archive/2014/10/why-germany-is-so-much-better-at-training-its-workers/381550/.

74. The US Chamber of Commerce claims to be "the world's largest business organization" and to represent more than 3 million businesses, www.uschamber.com/about-us-chamber-commerce. In 2017 the Chamber was the organization that spent the most lobbying the US Congress.

75. "Climate Change: The Path Forward," U.S. Chamber of Commerce, Sept. 27, 2019, www.uschamber.com/addressing-climate-change.

76. R. Meyer, "The Unprecedented Surge in Fear About Climate Change" [online], *Atlantic*, accessed Oct. 19, 2019, www.theatlantic.com/science/archive/2019/01/do-most-americans-believe-climate-change-polls-say-yes/580957/.

77. "Global Battery Alliance," *World Economic Forum*, www.weforum.org/projects/global-battery-alliance.

78. "Strengthening Global Food Systems," *World Economic Forum*, www.weforum.org/projects/strengthening-global-food-systems.

79. Anand Giridharadas, *Winners Take All: The Elite Charade of Changing the World* (New York: Alfred A. Knopf, 2018).

80. "ICC Launches New Tool to Promote Business Sustainability—ICC—International Chamber of Commerce," ICC, Jan. 19, 2017, https://iccwbo.org/media-wall/news-speeches/icc-launches-new-tool-to-promote-business-sustainability/.

81. Asbjørn Sonne Nørgaard, "Party Politics and the Organization of the Danish Welfare State, 1890–1920: The Bourgeois Roots of the Modern Welfare State," *Scandinavian Political Studies* 23, no. 3 (2000): 183–215.

82. Tim Worstall, "Denmark Does Not Have A $20 Minimum Wage, Try $11.70 Instead," *Forbes*, Aug. 13, 2015, www.forbes.com/sites/timworstall/2015/08/12/denmark-does-not-have-a-20-minimum-wage-try-11-70-instead/#17694b477814.

83. "Denmark Has OECD's Lowest Inequality," *Local*, May 21, 2015, www.thelocal.dk/20150521/denmark-has-lowest-inequality-among-oecd-nations. By contrast the gap in the United States is 18.8, with the richest 10 percent earning 18.8 times the poorest 10 percent.

84. Marc Sabatier Hvidkj, "How Does a Danish McDonald's Worker Make 20$/Hour, Without a Minimum Wage Law?" *Medium*, Jan. 30, 2019, https://medium.com/@marcsabatierhvidkjr/how-does-a-danish-mcdonalds-worker-make-20-hour-without-a-minimum-wage-law-ea8bcbaa870f.

85. James Edward Meade, "Mauritius: A Case Study in Malthusian Economics," *Economic Journal* 71, 283 (1961): 521–534.

86. The MLP campaigned for independence in alliance with the Comité d'Action Musulman (a Muslim party) and an outspokenly Hindu party, the Independence Forward Block.

87. Deborah Brautigam and Tania Diolle, "Coalitions, Capitalists and Credibility: Overcoming the Crisis of Confidence at Independence in Mauritius" (DLP Research Paper 4, 2009).

88. EPZs are areas that are designed to enable domestic manufacturers to compete internationally by freeing them of many local taxes and regulations. In 1962, a visiting World Bank mission exploring the possibility of setting some up in Mauritius had concluded: "A limited industrial expansion may take place mainly to serve domestic requirements. [But] a lack of domestic raw materials and of power supply, a shallow local market, great distances and relatively high labor costs set definite limits for industrial growth. Obviously, the conditions which helped Hong Kong or Puerto Rico to attain their present prominence do not exist in Mauritius."

89. Doing Business, "Training for Reform. Economy Profile Mauritius" (Washington, DC: World Bank Group, 2019), www.doingbusiness.org/content/dam/doingBusiness/country/m/mauritius/MUS.pdf.

90. "Mauritius," World Bank Data, https://data.worldbank.org/country/mauritius.

91. Lower Gini numbers mean less inequality. A society with a Gini coefficient of 0 would be perfectly equal. One with a coefficient of 100 would give all the income to a single person; "Countries Ranked by GINI Index (World Bank Estimate)," Index Mundi, www.indexmundi.com/facts/indicators/SI.POV.GINI/rankings.

92. *Human Development Indices and Indicators: 2018 Statistical Update* (Mauritius: UNDP, 2018), http://hdr.undp.org/sites/all/themes/hdr_theme/country-notes/MUS.pdf.

93. My sources for what follows are a personal interview with Daniella conducted in August 2018, my personal experience, and Leadership Now's website. I am on the Advisory Board of the organization.

Chapter 8. Pebbles in an Avalanche of Change

1. Hans Rosling, Ola Rosling, and Anna Rosling Rönnlund, *Factfulness: Ten Reasons We're Wrong About the World—and Why Things Are Better Than You Think*, 1st ed. (New York: Flatiron Books, 2018), 33.

2. K. Danziger, "Ideology and Utopia in South Africa: A Methodological Contribution to the Sociology of Knowledge," *British Journal of Sociology* 14, no. 1 (1963): 59–76.

3. Rosling et al. *Factfulness*, 53.

4. "Global Child Mortality: It Is Hard to Overestimate Both the Immensity of the Tragedy, and the Progress the World Has Made," *Our World in Data*, https://ourworldindata.org/child-mortality-globally.

5. Max Roser, "Economic Growth," *Our World in Data*, Nov. 24, 2013, https://ourworldindata.org/economic-growth; Max Roser et al., "World Population Growth,"

Our World in Data, May 9, 2013, https://ourworldindata.org/world-population-growth; "World Population by Year," *Worldometers*, www.worldometers.info/world-population /world-population-by-year/.

6. R. J. Reinhart, "Global Warming Age Gap: Younger Americans Most Worried," Gallup.com, Sept. 4, 2019, https://news.gallup.com/poll/234314/global-warming -age-gap-younger-americans-worried.aspx; Steven Pinker, *Enlightenment Now: The Case for Reason, Science, Humanism, and Progress* (New York: Viking, 2018).

7. Rosling et al., *Factfulness*, 60. More formally, per Wp is a "Watt-Peak" or a watt's worth of capacity under optimal conditions.

8. *Better Business Better World* (London: Business and Sustainable Development Commission, Jan. 2017), http://report.businesscommission.org/uploads/BetterBiz -BetterWorld_170215_012417.pdf.

9. "Renewable Energy Market Global Industry Analysis, Size, Share, Growth, Trends and Forecast 2019–2025," *Reuters*, Feb. 22, 2019, www.reuters.com/brandfeatures /venture-capital/article?id=85223.

10. "Renewables 2018 Global Status Report" (Paris: REN21 Secretariat).

11. Silvio Marcacci, "Renewable Energy Job Boom Creates Economic Opportunity as Coal Industry Slumps," *Forbes*, Apr. 22, 2019, www.forbes.com/sites /energyinnovation/2019/04/22/renewable-energy-job-boom-creating-economic -opportunity-as-coal-industry-slumps/#747b8f823665.

12. *Energy Efficiency Market Report 2018* (Paris: International Energy Agency [IEA], 2018), https://webstore.iea.org/download/direct/2369?fileName=Market_Report _Series_Energy_Efficiency_2018.pdf; OECD Publishing, *World Energy Outlook 2017* (Paris: Organization for Economic Cooperation and Development, 2017).

13. "Agriculture at a Crossroads," *Global Agriculture*, www.globalagriculture.org /report-topics/meat-and-animal-feed.html.

14. Deena Shanker, "Plant Based Foods Are Finding an Omnivorous Customer Base," Bloomberg.com, July 30, 2018, www.bloomberg.com/news/articles/2018 -07-30/plant-based-foods-are-finding-an-omnivorous-customer-base; Jesse Nichols and Eve Andrews, "How the Word 'Meat' Could Shape the Future of Protein," *Grist*, Jan. 18, 2019, https://grist.org/article/how-the-word-meat-could-shape-the -future-of-protein/; Janet Forgrieve, "Plant-Based Food Sales Continue to Grow by Double Digits, Fueled by Shift in Grocery Store Placement," *Forbes*, July 16, 2019, www.forbes.com/sites/janetforgrieve/2019/07/16/plant-based-food-sales-pick -up-the-pace-as-product-placement-shifts/#484fe50d4f75.

15. David Yaffe-Bellany, "The New Makers of Plant-Based Meat? Big Meat Companies," *New York Times*, Oct. 14, 2019, www.nytimes.com/2019/10/14/business/the-new -makers-of-plant-based-meat-big-meat-companies.html.

16. Hannah Ritchie and Max Roserm "Crop Yields," *Our World in Data*, Oct. 17, 2013, https://ourworldindata.org/yields-and-land-use-in-agriculture.

17. International Panel of Experts on Sustainable Food Systems (IPES-Food), "Breaking Away from Industrial Food and Farming Systems: Seven Case Studies of Agroecological Transition" (Oct. 2018); "Unlocking the Inclusive Growth Story of the 21st Century: Accelerating Climate Action in Urgent Times" (Washington, DC:

New Climate Economy, 2018), https://newclimateeconomy.report/2018/wp-content/uploads/sites/6/2018/09/NCE_2018_FULL-REPORT.pdf.; Technoserve, *Eyes in the Sky for African Agriculture, Water Resources, and Urban Planning*, Apr. 2018, www.technoserve.org/files/downloads/case-study_eyes-in-the-sky-for-african-agriculture-water-resources-and-urban-planning.pdf.

18. Food and Agriculture Organization (FAO), *The 10 Elements of Agroecology: Guiding the Transition to Sustainable Food and Agricultural Systems*, www.fao.org/3/i9037en/I9037EN.pdf; New Climate Economy, *Unlocking the Inclusive Growth Story* (2018).

19. For more on how to make a difference in your own life, and to connect with other readers of this book, please check out ReimaginingCapitalism.org.

20. Leor Hackel and Gregg Sparkman, "Actually, Your Personal Choices Do Make a Difference in Climate Change," *Slate Magazine*, Oct. 26, 2018, https://slate.com/technology/2018/10/carbon-footprint-climate-change-personal-action-collective-action.html.

21. Steve Westlake, "A Counter-Narrative to Carbon Supremacy: Do Leaders Who Give Up Flying Because of Climate Change Influence the Attitudes and Behaviour of Others?" SSRN 3283157 (2017).

22. Gregg Sparkman and Gregory M. Walton, "Dynamic Norms Promote Sustainable Behavior, Even If It Is Counternormative," *Psychological Science* 28, no. 11 (2017): 1663–1674.

23. Hackel and Sparkman, "Actually, Your Personal Choices Do Make a Difference."

24. Karen Asp, "WW Freestyle: Review for New Weight Watchers Plan," WebMD, Jan. 10, 2018, www.webmd.com/diet/a-z/weight-watchers-diet; John F. Kelly and Julie D. Yeterian, "The Role of Mutual-Help Groups in Extending the Framework of Treatment," *Alcohol Research & Health* 33, no. 4 (2011): 350, National Institute on Alcohol Abuse and Alcoholism; Dan Wagener, "What Is the Success Rate of AA?" *American Addiction Centers*, Oct. 28, 2019, https://americanaddictioncenters.org/rehab-guide/12-step/whats-the-success-rate-of-aa.

25. Ray Rothrick, "Rockefeller Family VC Funds Risky Fusion Energy Project," *Fusion 4 Freedom*, May 22, 2016, https://fusion4freedom.com/rockefeller-vc-funds-risky-fusion-project/.

26. Proforest is an NGO that works with large firms to help them transition to responsible agricultural sourcing. See www.proforest.net/en.

27. The account that follows draws from information available at www.mothersoutfront.org/.

28. "The Starfish Story," *City Year*, www.cityyear.org/about-us/culture-values/founding-stories/starfish-story, inspired by "The Star Thrower" a 16-page essay by Loren Eiseley, published in 1969 in *The Unexpected Universe*.

29. "Organizing Toolkit," Mothers Out Front, https://d3n8a8pro7vhmx.cloudfront.net/mothersoutfront/pages/1218/attachments/original/1494268006/MothersOutFront_toolkit-Section_1.pdf?1494268006.

30. Dennis Overbye, "John Huchra Dies at 61; Maps Altered Ideas on Universe," *New York Times*, Oct. 14, 2010, www.nytimes.com/2010/10/14/us/14huchra.html.

INDEX

PENGUIN PARTNERSHIPS